W9-CFZ-880

Student Edition

Eureka Math
Grade 5
Modules 1 & 2

Special thanks go to the Gordon A. Cain Center and to the Department of Mathematics at Louisiana State University for their support in the development of Eureka Math.

Published by the non-profit Great Minds

Copyright © 2015 Great Minds. No part of this work may be reproduced, sold, or commercialized, in whole or in part, without written permission from Great Minds. Non-commercial use is licensed pursuant to a Creative Commons Attribution-NonCommercial-ShareAlike 4.0 license; for more information, go to http://greatminds.net/maps/math/copyright. "Great Minds" and "Eureka Math" are registered trademarks of Great Minds.

Printed in the U.S.A.

This book may be purchased from the publisher at eureka-math.org

10 9 8 7 6 5 4 3

ISBN 978-1-63255-308-9

1,000,000	100,000	10,000	1,000	100	10	1	•	$\frac{1}{10}$	$\frac{1}{100}$	$\frac{1}{1000}$
Millions	Hundred Thousands	Ten Thousands	Thousands	Hundreds	Tens	Ones	•	Tenths	Hundredths	Thousandths
							•			
							•			
							•			

millions through thousandths place value chart

Lesson 1:   Reason concretely and pictorially using place value understanding to relate adjacent base ten units from millions to thousandths.

7

©2015 Great Minds. eureka-math.org
G5-M1-SE-B1-1.3.1-01.2016

This page intentionally left blank

Name _____   Date _____

1. Solve.

   a. 54,000 × 10 = _____

   b. 54,000 ÷ 10 = _____

   c. 8.7 × 10 = _____

   d. 8.7 ÷ 10 = _____

   e. 0.13 × 100 = _____

   f. 13 ÷ 1,000 = _____

   g. 3.12 × 1,000 = _____

   h. 4,031.2 ÷ 100 = _____

2. Find the products.

   a. 19,340 × 10   = _____

   b. 19,340 × 100   = _____

   c. 19,340 × 1,000 = _____

   d. Explain how you decided on the number of zeros in the products for (a), (b), and (c).

3. Find the quotients.

    a. 152 ÷ 10 = _____

    b. 152 ÷ 100 = _____

    c. 152 ÷ 1,000 = _____

    d. Explain how you decided where to place the decimal in the quotients for (a), (b), and (c).

4. Janice thinks that 20 hundredths is equivalent to 2 thousandths because 20 hundreds is equal to 2 thousands. Use words and a place value chart to correct Janice's error.

5. Canada has a population that is about $\frac{1}{10}$ as large as the United States. If Canada's population is about 32 million, about how many people live in the United States? Explain the number of zeros in your the answer.

10   Lesson 2:   Reason abstractly using place value understanding to relate adjacent
                                base ten units from millions to thousandths.

                                EUREKA
                                MATH

Name _____   Date _____

1. Solve.

    a. 36,000 × 10 = _____

    b. 36,000 ÷ 10 = _____

    c. 4.3 × 10 = _____

    d. 4.3 ÷ 10 = _____

    e. 2.4 x 100 = _____

    f. 24 ÷ 1,000 = _____

    g. 4.54 × 1,000 = _____

    h. 3,045.4 ÷ 100 = _____

2. Find the products.

    a. 14,560 × 10   = _____

    b. 14,560 × 100   = _____

    c. 14,560 × 1,000 = _____

    Explain how you decided on the number of zeros in the products for (a), (b), and (c).

EUREKA MATH

Lesson 2:   Reason abstractly using place value understanding to relate adjacent base ten units from millions to thousandths.

11

3. Find the quotients.

   a. 16.5 ÷ 10 = _____

   b. 16.5 ÷ 100 = _____

   c. Explain how you decided where to place the decimal in the quotients for (a) and (b).

4. Ted says that 3 tenths multiplied by 100 equals 300 thousandths. Is he correct? Use a place value chart to explain your answer.

5. Alaska has a land area of about 1,700,000 square kilometers. Florida has a land area $\frac{1}{10}$ the size of Alaska. What is the land area of Florida? Explain how you found your answer.

Lesson 2:   Reason abstractly using place value understanding to relate adjacent base ten units from millions to thousandths.

Name _____ Date _____

1. Write the following in exponential form (e.g., $100 = 10^2$).

    a. 10,000 = _____

    b. 1,000 = _____

    c. 10 × 10 = _____

    d. 100 × 100 = _____

    e. 1,000,000 = _____

    f. 1,000 × 1,000 = _____

2. Write the following in standard form (e.g., $5 × 10^2 = 500$).

    a. $9 × 10^3 =$ _____

    b. $39 × 10^4 =$ _____

    c. $7,200 ÷ 10^2 =$ _____

    d. $7,200,000 ÷ 10^3 =$ _____

    e. $4.025 × 10^3 =$ _____

    f. $40.25 × 10^4 =$ _____

    g. $72.5 ÷ 10^2 =$ _____

    h. $7.2 ÷ 10^2 =$ _____

3. Think about the answers to Problem 2(a–d). Explain the pattern used to find an answer when you multiply or divide a whole number by a power of 10.

4. Think about the answers to Problem 2(e–h). Explain the pattern used to place the decimal in the answer when you multiply or divide a decimal by a power of 10.

EUREKA
MATH™

Lesson 3:   Use exponents to name place value units, and explain patterns in the
placement of the decimal point.

13

5. Complete the patterns.

a. 0.03   0.3   _____   30   _____   _____

b. 6,500,000   65,000   _____   6.5   _____

c. _____   9,430   _____   94.3   9.43   _____

d. 999   9990   99,900   _____   _____   _____

e. _____   7.5   750   75,000   _____   _____

f. Explain how you found the unknown numbers in set (b). Be sure to include your reasoning about the number of zeros in your numbers and how you placed the decimal.

g. Explain how you found the unknown numbers in set (d). Be sure to include your reasoning about the number of zeros in your numbers and how you placed the decimal.

6. Shaunnie and Marlon missed the lesson on exponents. Shaunnie incorrectly wrote $10^5 = 50$ on her paper, and Marlon incorrectly wrote $2.5 \times 10^2 = 2.500$ on his paper.

a. What mistake has Shaunnie made? Explain using words, numbers, or pictures why her thinking is incorrect and what she needs to do to correct her answer.

b. What mistake has Marlon made? Explain using words, numbers, or pictures why his thinking is incorrect and what he needs to do to correct his answer.

Lesson 3:   Use exponents to name place value units, and explain patterns in the placement of the decimal point.

EUREKA
MATH™

Name _____   Date _____

1. Write the following in exponential form (e.g., $100 = 10^2$).

    a. 1000 = _____   d. 100 × 10 = _____

    b. 10 × 10 = _____   e. 1,000,000 = _____

    c. 100,000 = _____   f. 10,000 × 10 = _____

2. Write the following in standard form (e.g., $4 × 10^2 = 400$).

    a. $4 × 10^3$ = _____   e. $6.072 × 10^3$ = _____

    b. $64 × 10^4$ = _____   f. $60.72 × 10^4$ = _____

    c. $5,300 ÷ 10^2$ = _____   g. $948 ÷ 10^3$ = _____

    d. $5,300,000 ÷ 10^3$ = _____ h. $9.4 ÷ 10^2$ = _____

3. Complete the patterns.

    a. 0.02   0.2   _____   20   _____   _____

    b. 3,400,000   34,000   _____   3.4   _____

    c. _____   8,570   _____   85.7   8.57   _____

    d. 444   4440   44,400   _____   _____   _____

    e. _____   9.5   950   95,000   _____   _____

EUREKA
MATH™

Lesson 3:   Use exponents to name place value units, and explain patterns in the
             placement of the decimal point.

15

4. After a lesson on exponents, Tia went home and said to her mom, "I learned that 10^4 is the same as 40,000." She has made a mistake in her thinking. Use words, numbers, or a place value chart to help Tia correct her mistake.

5. Solve $247 \div 10^2$ and 247×10^2.

    a. What is different about the two answers? Use words, numbers, or pictures to explain how the digits shift.

    b. Based on the answers from the pair of expressions above, solve $247 \div 10^3$ and 247×10^3.

     Lesson 3:   Use exponents to name place value units, and explain patterns in the placement of the decimal point.

    EUREKA MATH™

10		
$10 \times$ ___		

powers of 10 chart

Lesson 3:   Use exponents to name place value units, and explain patterns in the placement of the decimal point.

17

This page intentionally left blank

Name _____ Date _____

1. Convert and write an equation with an exponent. Use your meter strip when it helps you.

   a. 3 meters to centimeters   3 m = 300 cm $\underline{\quad 3 \times 10^2 = 300 \quad}$

   b. 105 centimeters to meters   105 cm = _____ m   _____

   c. 1.68 meters to centimeters   _____ m = _____ cm   _____

   d. 80 centimeters to meters   _____ cm = _____ m   _____

   e. 9.2 meters to centimeters   _____ m = _____ cm   _____

   f. 4 centimeters to meters   _____ cm = _____ m   _____

   g. In the space below, list the letters of the problems where larger units are converted to smaller units.

2. Convert using an equation with an exponent. Use your meter strip when it helps you.

   a. 3 meters to millimeters   _____ m = _____ mm   _____

   b. 1.2 meters to millimeters   _____ m = _____ mm   _____

   c. 1,020 millimeters to meters   _____ mm = _____ m   _____

   d. 97 millimeters to meters   _____ mm = _____ m   _____

   e. 7.28 meters to millimeters   _____ m = _____ mm   _____

   f. 4 millimeters to meters   _____ mm = _____ m   _____

   g. In the space below, list the letters of the problems where smaller units are converted to larger units.

EUREKA MATH™

Lesson 4: Use exponents to denote powers of 10 with application to metric conversions.

19

3. Read each aloud as you write the equivalent measures. Write an equation with an exponent you might use to convert.

   a. 3.512 m = _____ mm $3.512 \times 10^3 = 3{,}512$

   b. 8 cm = _____ m   _____

   c. 42 mm = _____ m   _____

   d. 0.05 m = _____ mm   _____

   e. 0.002 m = _____ cm   _____

4. The length of the bar for a high jump competition must always be 4.75 m. Express this measurement in millimeters. Explain your thinking. Include an equation with an exponent in your explanation.

5. A honey bee's length measures 1 cm. Express this measurement in meters. Explain your thinking. Include an equation with an exponent in your explanation.

6. Explain why converting from meters to centimeters uses a different exponent than converting from meters to millimeters.

Lesson 4:   Use exponents to denote powers of 10 with application to metric conversions.

©2015 Great Minds. eureka-math.org
G5-M1-SE-B1-1.3.1-01.2016

EUREKA MATH™

Name _____   Date _____

1. Convert and write an equation with an exponent. Use your meter strip when it helps you.

    a. 2 meters to centimeters   2m = 200 cm   _____ $2 \times 10^2 = 200$ _____

    b. 108 centimeters to meters   108 cm = _____ m   _____

    c. 2.49 meters to centimeters   _____ m = _____ cm _____

    d. 50 centimeters to meters   _____ cm = _____ m _____

    e. 6.3 meters to centimeters   _____ m = _____ cm _____

    f. 7 centimeters to meters   _____ cm = _____ m _____

    g. In the space below, list the letters of the problems where smaller units are converted to larger units.

2. Convert using an equation with an exponent. Use your meter strip when it helps you.

    a. 4 meters to millimeters   _____ m = _____ mm   _____

    b. 1.7 meters to millimeters   _____ m = _____ mm   _____

    c. 1,050 millimeters to meters   _____ mm = _____ m   _____

    d. 65 millimeters to meters   _____ mm = _____ m   _____

    e. 4.92 meters to millimeters   _____ m = _____ mm   _____

    f. 3 millimeters to meters   _____ mm = _____ m   _____

    g. In the space below, list the letters of the problems where larger units are converted to smaller units.

EUREKA
MATH™

Lesson 4:   Use exponents to denote powers of 10 with application to metric
             conversions.

21

3. Read each aloud as you write the equivalent measures. Write an equation with an exponent you might use to convert.

a. 2.638 m   = _____ mm   _____$2.638 \times 10^3 = 2,638$_____

b. 7 cm   = _____ m   _____

c. 39 mm   = _____ m   _____

d. 0.08 m   = _____ mm   _____

e. 0.005 m   = _____ cm   _____

4. Yi Ting's height is 1.49 m. Express this measurement in millimeters. Explain your thinking. Include an equation with an exponent in your explanation.

5. A ladybug's length measures 2 cm. Express this measurement in meters. Explain your thinking. Include an equation with an exponent in your explanation.

6. The length of a sticky note measures 77 millimeters. Express this length in meters. Explain your thinking. Include an equation with an exponent in your explanation.

Lesson 4:   Use exponents to denote powers of 10 with application to metric conversions.

EUREKA MATH

Name _____ Date _____

1. Express as decimal numerals. The first one is done for you.

a. Four thousandths	0.004
b. Twenty-four thousandths	
c. One and three hundred twenty-four thousandths	
d. Six hundred eight thousandths	
e. Six hundred and eight thousandths	
f. $\frac{46}{1000}$	
g. $3\frac{946}{1000}$	
h. $200\frac{904}{1000}$	

2. Express each of the following values in words.

    a. 0.005   _____

    b. 11.037   _____

    c. 403.608   _____

3. Write the number on a place value chart. Then, write it in expanded form using fractions or decimals to express the decimal place value units. The first one is done for you.

    a. 35.827

Tens	Ones		Tenths	Hundredths	Thousandths
3	5	●	8	2	7

$$35.827 = 3 \times 10 + 5 \times 1 + 8 \times \left(\frac{1}{10}\right) + 2 \times \left(\frac{1}{100}\right) + 7 \times \left(\frac{1}{1000}\right) \ \ or$$
$$= 3 \times 10 + 5 \times 1 + 8 \times 0.1 + 2 \times 0.01 + 7 \times 0.001$$

Lesson 5: Name decimal fractions in expanded, unit, and word forms by applying place value reasoning.

23

b. 0.249

c. 57.281

4. Write a decimal for each of the following. Use a place value chart to help, if necessary.

a. $7 \times 10 + 4 \times 1 + 6 \times \left(\frac{1}{10}\right) + 9 \times \left(\frac{1}{100}\right) + 2 \times \left(\frac{1}{1000}\right)$

b. $5 \times 100 + 3 \times 10 + 8 \times 0.1 + 9 \times 0.001$

c. $4 \times 1,000 + 2 \times 100 + 7 \times 1 + 3 \times \left(\frac{1}{100}\right) + 4 \times \left(\frac{1}{1000}\right)$

5. Mr. Pham wrote 2.619 on the board. Christy says it is two and six hundred nineteen thousandths. Amy says it is 2 ones 6 tenths 1 hundredth 9 thousandths. Who is right? Use words and numbers to explain your answer.

Lesson 5: Name decimal fractions in expanded, unit, and word forms by applying place value reasoning.

EUREKA MATH™

Name _____   Date _____

1. Express as decimal numerals. The first one is done for you.

a. Five thousandths	0.005
b. Thirty-five thousandths	
c. Nine and two hundred thirty-five thousandths	
d. Eight hundred and five thousandths	
e. $\frac{8}{1000}$	
f. $\frac{28}{1000}$	
g. $7\frac{528}{1000}$	
h. $300\frac{502}{1000}$	

2. Express each of the following values in words.

a. 0.008 _____

b. 15.062 _____

c. 607.409 _____

3. Write the number on a place value chart. Then, write it in expanded form using fractions or decimals to express the decimal place value units. The first one is done for you.

a. 27.346

Tens	Ones		Tenths	Hundredths	Thousandths
2	7	●	3	4	6

$27.346 = 2 \times 10 + 7 \times 1 + 3 \times \left(\frac{1}{10}\right) + 4 \times \left(\frac{1}{100}\right) + 6 \times \left(\frac{1}{1000}\right)$ *or*

$27.346 = 2 \times 10 + 7 \times 1 + 3 \times 0.1 + 4 \times 0.01 + 6 \times 0.001$

Lesson 5:   Name decimal fractions in expanded, unit, and word forms by applying place value reasoning.

25

b. 0.362

c. 49.564

4. Write a decimal for each of the following. Use a place value chart to help, if necessary.

   a. $3 \times 10 + 5 \times 1 + 2 \times \left(\frac{1}{10}\right) + 7 \times \left(\frac{1}{100}\right) + 6 \times \left(\frac{1}{1000}\right)$

   b. $9 \times 100 + 2 \times 10 + 3 \times 0.1 + 7 \times 0.001$

   c. $5 \times 1000 + 4 \times 100 + 8 \times 1 + 6 \times \left(\frac{1}{100}\right) + 5 \times \left(\frac{1}{1000}\right)$

5. At the beginning of a lesson, a piece of chalk is 4.875 inches long. At the end of the lesson, it is 3.125 inches long. Write the two amounts in expanded form using fractions.

   a. At the beginning of the lesson:

   b. At the end of the lesson:

6. Mrs. Herman asked the class to write an expanded form for 412.638. Nancy wrote the expanded form using fractions, and Charles wrote the expanded form using decimals. Write their responses.

EUREKA MATH™

Thousandths	Hundredths	Tenths	Ones	Tens	Hundreds	Thousands
			●			

thousands through thousandths place value chart

Lesson 5:   Name decimal fractions in expanded, unit, and word forms by applying place value reasoning.

27

This page intentionally left blank

Name _____   Date _____

1. Show the numbers on the place value chart using digits. Use >, <, or = to compare. Explain your thinking in the space to the right.

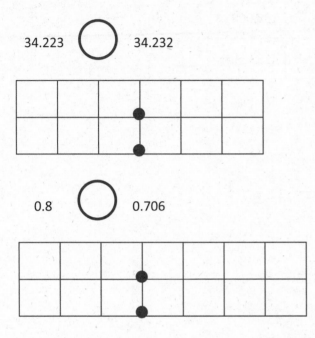

34.223 ◯ 34.232

0.8 ◯ 0.706

2. Use >, <, or = to compare the following. Use a place value chart to help, if necessary.

a. 16.3	◯	16.4
b. 0.83	◯	$\frac{83}{100}$
c. $\frac{205}{1000}$	◯	0.205
d. 95.580	◯	95.58
e. 9.1	◯	9.099
f. 8.3	◯	83 tenths
g. 5.8	◯	Fifty-eight hundredths

EUREKA MATH

Lesson 6: Compare decimal fractions to the thousandths using like units, and express comparisons with >, <, =.

29

h. Thirty-six and nine thousandths	◯	4 tens
i. 202 hundredths	◯	2 hundreds and 2 thousandths
j. One hundred fifty-eight thousandths	◯	158,000
k. 4.15	◯	415 tenths

3. Arrange the numbers in increasing order.

a. 3.049   3.059   3.05   3.04

_____

b. 182.205   182.05   182.105   182.025

_____

4. Arrange the numbers in decreasing order.

a. 7.608   7.68   7.6   7.068

_____

b. 439.216   439.126   439.612   439.261

_____

Lesson 6:   Compare decimal fractions to the thousandths using like units, and express comparisons with >, <, =.

EUREKA MATH™

5. Lance measured 0.485 liter of water. Angel measured 0.5 liter of water. Lance said, "My beaker has more water than yours because my number has three decimal places and yours only has one." Is Lance correct? Use words and numbers to explain your answer.

6. Dr. Hong prescribed 0.019 liter more medicine than Dr. Tannenbaum. Dr. Evans prescribed 0.02 less than Dr. Hong. Who prescribed the most medicine? Who prescribed the least?

Lesson 6:   Compare decimal fractions to the thousandths using like units, and express comparisons with >, <, =.

31

Name _____   Date _____

1. Use >, <, or = to compare the following.

a. 16.45	◯	16.454
b. 0.83	◯	$\frac{83}{100}$
c. $\frac{205}{1000}$	◯	0.205
d. 95.045	◯	95.545
e. 419.10	◯	419.099
f. Five ones and eight tenths	◯	Fifty-eight tenths
g. Thirty-six and nine thousandths	◯	Four tens
h. One hundred four and twelve hundredths	◯	One hundred four and two thousandths
i. One hundred fifty-eight thousandth	◯	0.58
j. 703.005	◯	Seven hundred three and five hundredths

2. Arrange the numbers in increasing order.

    a.   8.08   8.081   8.09   8.008

    _____

    b.   14.204   14.200   14.240   14.210

    _____

Lesson 6:   Compare decimal fractions to the thousandths using like units, and
express comparisons with >, <, =.

©2015 Great Minds. eureka-math.org
G5-M1-SE-B1-1.3.1-01.2016

3. Arrange the numbers in decreasing order.

    a.   8.508   8.58   7.5   7.058

    _____

    b.   439.216   439.126   439.612   439.261

    _____

4. James measured his hand. It was 0.17 meter. Jennifer measured her hand. It was 0.165 meter. Whose hand is bigger? How do you know?

5. In a paper airplane contest, Marcel's plane travels 3.345 meters. Salvador's plane travels 3.35 meters. Jennifer's plane travels 3.3 meters. Based on the measurements, whose plane traveled the farthest distance? Whose plane traveled the shortest distance? Explain your reasoning using a place value chart.

This page intentionally left blank

Name _____   Date _____

Fill in the table, and then round to the given place. Label the number lines to show your work. Circle the rounded number.

1. 3.1

a. Hundredths   b. Tenths   c. Tens

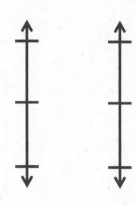

Tens	Ones	Tenths	Hundredths	Thousandths

2. 115.376

a. Hundredths   b. Ones   c. Tens

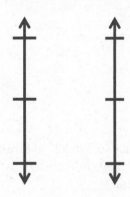

Tens	Ones	Tenths	Hundredths	Thousandths

EUREKA
MATH

Lesson 7:   Round a given decimal to any place using place value understanding
and the vertical number line.

35

©2015 Great Minds. eureka-math.org
G5-M1-SE-B1-1.3.1-01.2016

3. 0.994

Tens	Ones	Tenths	Hundredths	Thousandths

a. Hundredths

b. Tenths

c. Ones

d. Tens

4. For open international competition, the throwing circle in the men's shot put must have a diameter of 2.135 meters. Round this number to the nearest hundredth. Use a number line to show your work.

5. Jen's pedometer said she walked 2.549 miles. She rounded her distance to 3 miles. Her brother rounded her distance to 2.5 miles. When they argued about it, their mom said they were both right. Explain how that could be true. Use number lines and words to explain your reasoning.

Lesson 7: Round a given decimal to any place using place value understanding and the vertical number line.

EUREKA MATH

Name _____   Date _____

Fill in the table, and then round to the given place. Label the number lines to show your work. Circle the rounded number.

1. 4.3

    a. Hundredths   b. Tenths   c. Ones

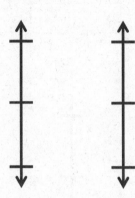

Tens	Ones	Tenths	Hundredths	Thousandths
	●			

2. 225.286

    a. Hundredths   b. Ones   c. Tens

Tens	Ones	Tenths	Hundredths	Thousandths
	●			

EUREKA MATH

Lesson 7:   Round a given decimal to any place using place value understanding and the vertical number line.

37

3. 8.984

Tens	Ones	Tenths	Hundredths	Thousandths

a. Hundredths   b. Tenths   c. Ones   d. Tens

4. On a Major League Baseball diamond, the distance from the pitcher's mound to home plate is 18.386 meters.

   a. Round this number to the nearest hundredth of a meter. Use a number line to show your work.

   b. How many centimeters is it from the pitcher's mound to home plate?

5. Jules reads that 1 pint is equivalent to 0.473 liters. He asks his teacher how many liters there are in a pint. His teacher responds that there are about 0.47 liters in a pint. He asks his parents, and they say there are about 0.5 liters in a pint. Jules says they are both correct. How can that be true? Explain your answer.

Lesson 7:   Round a given decimal to any place using place value understanding and the vertical number line.

EUREKA
MATH™

		Thousandths			
		Hundredths			
		Tenths			
		•			
		Ones			
		Tens			
		Hundreds			

hundreds to thousandths place value chart

Lesson 7:   Round a given decimal to any place using place value understanding
            and the vertical number line.

39

This page intentionally left blank

Name _____   Date _____

1. Write the decomposition that helps you, and then round to the given place value. Draw number lines to explain your thinking. Circle the rounded value on each number line.

   a. Round 32.697 to the nearest tenth, hundredth, and one.

   b. Round 141.999 to the nearest tenth, hundredth, ten, and hundred.

2. A root beer factory produces 132,554 cases in 100 days. About how many cases does the factory produce in 1 day? Round your answer to the nearest tenth of a case. Show your thinking on the number line.

Lesson 8:   Round a given decimal to any place using place value understanding and the vertical number line.

41

3. A decimal number has two digits to the right of its decimal point. If we round it to the nearest tenth, the result is 13.7.

    a. What is the maximum possible value of this number? Use words and the number line to explain your reasoning. Include the midpoint on your number line.

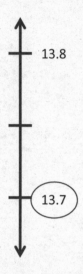

    b. What is the minimum possible value of this decimal? Use words and the number line to explain your reasoning. Include the midpoint on your number line.

Lesson 8:   Round a given decimal to any place using place value understanding
                      and the vertical number line.

©2015 Great Minds. eureka-math.org
G5-M1-SE-B1-1.3.1-01.2016

Name _____   Date _____

1. Write the decomposition that helps you, and then round to the given place value. Draw number lines to explain your thinking. Circle the rounded value on each number line.

   a. 43.586 to the nearest tenth, hundredth, and one.

   b. 243.875 to nearest tenth, hundredth, ten, and hundred.

2. A trip from New York City to Seattle is 2,852.1 miles. A family wants to make the drive in 10 days, driving the same number of miles each day. About how many miles will they drive each day? Round your answer to the nearest tenth of a mile.

Lesson 8:   Round a given decimal to any place using place value understanding   **43**
                and the vertical number line.

©2015 Great Minds. eureka-math.org
G5-M1-SE-B1-1.3.1-01.2016

3. A decimal number has two digits to the right of its decimal point. If we round it to the nearest tenth, the result is 18.6.

   a. What is the maximum possible value of this number? Use words and the number line to explain your reasoning. Include the midpoint on your number line.

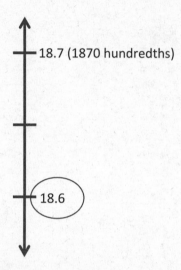

18.7 (1870 hundredths)

18.6

   b. What is the minimum possible value of this decimal? Use words, pictures, or numbers to explain your reasoning.

18.6

18.5

EUREKA
MATH™

Name _____   Date _____

1. Solve, and then write the sum in standard form. Use a place value chart if necessary.

a.   1 tenth + 2 tenths = _____ tenths = _____

b.   14 tenths + 9 tenths = _____ tenths = _____ one(s) _____ tenth(s) = _____

c.   1 hundredth + 2 hundredths = _____ hundredths = _____

d.   27 hundredths + 5 hundredths = _____ hundredths = _____ tenths _____ hundredths = _____

e.   1 thousandth + 2 thousandths = _____ thousandths = _____

f.   35 thousandths + 8 thousandths = _____ thousandths = _____ hundredths _____ thousandths = _____

g.   6 tenths + 3 thousandths = _____ thousandths = _____

h.   7 ones 2 tenths + 4 tenths = _____ tenths = _____

i.   2 thousandths + 9 ones 5 thousandths = _____ thousandths = _____

2. Solve using the standard algorithm.

a. 0.3 + 0.82 = _____	b. 1.03 + 0.08 = _____
c. 7.3 + 2.8 = _____	d. 57.03 + 2.08 = _____

e. 62.573 + 4.328 = _____	f. 85.703 + 12.197 = _____

3. Van Cortlandt Park's walking trail is 1.02 km longer than Marine Park's. Central Park's walking trail is 0.242 km longer than Van Cortlandt's.

a.   Fill in the missing information in the chart below.

New York City Walking Trails	
Central Park	_____ km
Marine Park	1.28 km
Van Cortlandt Park	_____ km

b.   If a tourist walked all 3 trails in a day, how many kilometers would he or she have walked?

4. Meyer has 0.64 GB of space remaining on his iPod. He wants to download a pedometer app (0.24 GB), a photo app (0.403 GB), and a math app (0.3 GB). Which combinations of apps can he download? Explain your thinking.

Lesson 9:   Add decimals using place value strategies, and relate those strategies to a written method.

EUREKA MATH

©2015 Great Minds. eureka-math.org
G5-M1-SE-B1-1.3.1-01.2016

Name _____   Date _____

1. Solve.

    a. 3 tenths + 4 tenths = _____ tenths

    b. 12 tenths + 9 tenths = _____ tenths = _____ one(s) _____ tenth(s)

    c. 3 hundredths + 4 hundredths = _____ hundredths

    d. 27 hundredths + 7 hundredths = _____ hundredths = _____ tenths _____ hundredths

    e. 4 thousandths + 3 thousandths = _____ thousandths

    f. 39 thousandths + 5 thousandths = ____ thousandths = ____ hundredths ____ thousandths

    g. 5 tenths + 7 thousandths = _____ thousandths

    h. 4 ones 4 tenths + 4 tenths = _____ tenths

    i. 8 thousandths + 6 ones 8 thousandths = _____ thousandths

2. Solve using the standard algorithm.

a. 0.4 + 0.7 = _____	b. 2.04 + 0.07 = _____
c. 6.4 + 3.7 = _____	d. 56.04 + 3.07 = _____

| e. 72.564 + 5.137 = _____ | f. 75.604 + 22.296 = _____ |
| | |

3. Walkway Over the Hudson, a bridge that crosses the Hudson River in Poughkeepsie, is 2.063 kilometers long. Anping Bridge, which was built in China 850 years ago, is 2.07 kilometers long.

   a. What is the total span of both bridges? Show your thinking.

   b. Leah likes to walk her dog on the Walkway Over the Hudson. If she walks across and back, how far will she and her dog walk?

4. For his parents' anniversary, Danny spends $5.87 on a photo. He also buys a balloon for $2.49 and a box of strawberries for $4.50. How much money does he spend all together?

Lesson 9: Add decimals using place value strategies, and relate those strategies to a written method.

EUREKA MATH™

Name _____ Date _____

1. Subtract, writing the difference in standard form. You may use a place value chart to solve.

    a. 5 tenths – 2 tenths = _____ tenths = _____

    b. 5 ones 9 thousandths – 2 ones = _____ ones _____ thousandths = _____

    c. 7 hundreds 8 hundredths – 4 hundredths = _____ hundreds _____ hundredths = _____

    d. 37 thousandths – 16 thousandths = _____ thousandths = _____

2. Solve using the standard algorithm.

a. 1.4 – 0.7 = _____	b. 91.49 – 0.7 = _____	c. 191.49 – 10.72 = _____
d. 7.148 – 0.07 = _____	e. 60.91 – 2.856 = _____	f. 361.31 – 2.841 = _____

EUREKA MATH

Lesson 10: Subtract decimals using place value strategies, and relate those strategies to a written method.

49

3. Solve.

a. 10 tens – 1 ten 1 tenth	b. 3 – 22 tenths	c. 37 tenths – 1 one 2 tenths
d. 8 ones 9 hundredths – 3.4	e. 5.622 – 3 hundredths	f. 2 ones 4 tenths – 0.59

4. Mrs. Fan wrote *5 tenths minus 3 hundredths* on the board. Michael said the answer is 2 tenths because 5 minus 3 is 2. Is he correct? Explain.

5. A pen costs $2.09. It costs $0.45 less than a marker. Ken paid for one pen and one marker with a five-dollar bill. Use a tape diagram with calculations to determine his change.

Lesson 10: Subtract decimals using place value strategies, and relate those strategies to a written method.

EUREKA MATH™

Name _____   Date _____

1. Subtract. You may use a place value chart.

   a. 9 tenths – 3 tenths = _____ tenths

   b. 9 ones 2 thousandths – 3 ones = _____ones _____ thousandths

   c. 4 hundreds 6 hundredths – 3 hundredths = _____ hundreds _____ hundredths

   d. 56 thousandths – 23 thousandths = _____ thousandths = _____ hundredths _____ thousandths

2. Solve using the standard algorithm.

a. 1.8 – 0.9 = _____	b. 41.84 – 0.9 = _____	c. 341.84 – 21.92 = _____
d. 5.182 – 0.09 = _____	e. 50.416 – 4.25 = _____	f. 741 – 3.91 = _____

3. Solve.

a. 30 tens – 3 tens 3 tenths	b. 5 – 16 tenths	c. 24 tenths – 1 one 3 tenths
d. 6 ones 7 hundredths – 2.3	e. 8.246 – 5 hundredths	f. 5 ones 3 tenths – 0.53

4. Mr. House wrote *8 tenths minus 5 hundredths* on the board. Maggie said the answer is 3 hundredths because 8 minus 5 is 3. Is she correct? Explain.

5. A clipboard costs $2.23. It costs $0.58 more than a notebook. Lisa bought two clipboards and one notebook. She paid with a ten-dollar bill. How much change does Lisa get? Use a tape diagram to show your thinking.

Lesson 10:   Subtract decimals using place value strategies, and relate those strategies to a written method.

EUREKA MATH™

©2015 Great Minds. eureka-math.org
G5-M1-SE-B1-1.3.1-01.2016

Name _____   Date _____

1. Solve by drawing disks on a place value chart. Write an equation, and express the product in standard form.

    a. 3 copies of 2 tenths   b. 5 groups of 2 hundredths

    c. 3 times 6 tenths   d. 6 times 4 hundredths

    e. 5 times as much as 7 tenths   f. 4 thousandths times 3

2. Draw a model similar to the one pictured below for Parts (b), (c), and (d). Find the sum of the partial products to evaluate each expression.

    a. 7×3.12

	3 ones	+	1 tenth	+	2 hundredths
7	7×3 ones		7×1 tenth		7×2 hundredths

   _____ + _____ + 0.14 = _____

    b. 6×4.25

Lesson 11:   Multiply a decimal fraction by single-digit whole numbers, relate to a written method through application of the area model and place value understanding, and explain the reasoning used.   **53**

©2015 Great Minds. eureka-math.org
G5-M1-SE-B1-1.3.1-01.2016

c.   3 copies of 4.65

d.   4 times as much as 20.075

3.   Miles incorrectly gave the product of 7×2.6 as 14.42. Use a place value chart or an area model to help Miles understand his mistake.

4.   Mrs. Zamir wants to buy 8 protractors and some erasers for her classroom. She has $30. If protractors cost $2.65 each, how much will Mrs. Zamir have left to buy erasers?

Lesson 11:   Multiply a decimal fraction by single-digit whole numbers, relate to a written method through application of the area model and place value understanding, and explain the reasoning used.

©2015 Great Minds. eureka-math.org
G5-M1-SE-B1-1.3.1-01.2016

EUREKA
MATH™

Name _____ Date _____

1. Solve by drawing disks on a place value chart. Write an equation, and express the product in standard form.

   a. 2 copies of 4 tenths

   b. 4 groups of 5 hundredths

   c. 4 times 7 tenths

   d. 3 times 5 hundredths

   e. 9 times as much as 7 tenths

   f. 6 thousandths times 8

2. Draw a model similar to the one pictured below. Find the sum of the partial products to evaluate each expression.

   a. 4×6.79

	6 ones +	7 tenths +	9 hundredths
4	4 × 6 ones	4 × 7 tenths	4 × 9 hundredths

   _____ + _____ + _____ = _____

Lesson 11:   Multiply a decimal fraction by single-digit whole numbers, relate to a written method through application of the area model and place value understanding, and explain the reasoning used.

55

EUREKA MATH™

©2015 Great Minds. eureka-math.org
G5-M1-SE-B1-1.3.1-01.2016

b. 6×7.49

c. 9 copies of 3.65

d. 3 times 20.175

3. Leanne multiplied 8×4.3 and got 32.24. Is Leanne correct? Use an area model to explain your answer.

4. Anna buys groceries for her family. Hamburger meat is \$3.38 per pound, sweet potatoes are \$0.79 each, and hamburger rolls are \$2.30 a bag. If Anna buys 3 pounds of meat, 5 sweet potatoes, and 1 bag of hamburger rolls, what will she pay in all for the groceries?

Lesson 11: Multiply a decimal fraction by single-digit whole numbers, relate to a written method through application of the area model and place value understanding, and explain the reasoning used.

EUREKA
MATH™

Name _____ Date _____

1. Choose the reasonable product for each expression. Explain your reasoning in the spaces below using words, pictures, or numbers.

   a. 2.5 × 4   0.1   1   10   100

   _____

   b. 3.14 × 7   2198   219.8   21.98   2.198

   _____

   c. 8 × 6.022   4.8176   48.176   481.76   4817.6

   _____

   d. 9 × 5.48   493.2   49.32   4.932   0.4932

EUREKA MATH

Lesson 12: Multiply a decimal fraction by single-digit whole numbers, including using estimation to confirm the placement of the decimal point.

57

©2015 Great Minds. eureka-math.org
G5-M1-SE-B1-1.3.1-01.2016

2. Pedro is building a spice rack with 4 shelves that are each 0.55 meter long. At the hardware store, Pedro finds that he can only buy the shelving in whole meter lengths. Exactly how many meters of shelving does Pedro need? Since he can only buy whole-number lengths, how many meters of shelving should he buy? Justify your thinking.

3. Marcel rides his bicycle to school and back on Tuesdays and Thursdays. He lives 3.62 kilometers away from school. Marcel's gym teacher wants to know about how many kilometers he bikes in a week. Marcel's math teacher wants to know exactly how many kilometers he bikes in a week. What should Marcel tell each teacher? Show your work.

4. The poetry club had its first bake sale, and they made $79.35. The club members are planning to have 4 more bake sales. Leslie said, "If we make the same amount at each bake sale, we'll earn $3,967.50." Peggy said, "No way, Leslie! We'll earn $396.75 after five bake sales." Use estimation to help Peggy explain why Leslie's reasoning is inaccurate. Show your reasoning using words, numbers, or pictures.

Lesson 12: Multiply a decimal fraction by single-digit whole numbers, including using estimation to confirm the placement of the decimal point.

EUREKA MATH

©2015 Great Minds. eureka-math.org
G5-M1-SE-B1-1.3.1-01.2016

Name _____   Date _____

1. Choose the reasonable product for each expression. Explain your thinking in the spaces below using words, pictures, or numbers.

   a.   2.1 × 3   0.63   6.3   63   630

   _____

   b.   4.27 × 6   2562   256.2   25.62   2.562

   _____

   c.   7 × 6.053   4237.1   423.71   42.371   4.2371

   _____

   d.   9 × 4.82   4.338   43.38   433.8   4338

©2015 Great Minds. eureka-math.org
G5-M1-SE-B1-1.3.1-01.2016

2. Yi Ting weighs 8.3 kg. Her older brother is 4 times as heavy as Yi Ting. How much does her older brother weigh in kilograms?

3. Tim is painting his storage shed. He buys 4 gallons of white paint and 3 gallons of blue paint. Each gallon of white paint costs $15.72, and each gallon of blue paint is $21.87. How much will Tim spend in all on paint?

4. Ribbon is sold at 3 yards for $6.33. Jackie bought 24 yards of ribbon for a project. How much did she pay?

Lesson 12:   Multiply a decimal fraction by single-digit whole numbers, including using estimation to confirm the placement of the decimal point.

Name _____   Date _____

1. Complete the sentences with the correct number of units, and then complete the equation.

    a. 4 groups of _____ tenths is 1.6.   1.6 ÷ 4 = _____

    b. 8 groups of _____ hundredths is 0.32.   0.32 ÷ 8 = _____

    c. 7 groups of _____ thousandths is 0.084.   0.084 ÷ 7 = _____

    d. 5 groups of _____ tenths is 2.0.   2.0 ÷ 5 = _____

2. Complete the number sentence. Express the quotient in units and then in standard form.

    a. 4.2 ÷ 7 = _____ tenths ÷ 7 = _____ tenths = _____

    b. 2.64 ÷ 2 = _____ ones ÷ 2 + _____ hundredths ÷ 2

        = _____ ones + _____ hundredths

        = _____

    c. 12.64 ÷ 2 = _____ ones ÷ 2 + _____ hundredths ÷ 2

        = _____ ones + _____ hundredths

        = _____

    d. 4.26 ÷ 6 = _____ tenths ÷ 6 + _____ hundredths ÷ 6

        = _____

        = _____

Lesson 13:   Divide decimals by single-digit whole numbers involving easily
identifiable multiples using place value understanding and relate
to a written method.

61

e. $4.236 \div 6 =$ _____

   $=$ _____

   $=$ _____

3. Find the quotients. Then, use words, numbers, or pictures to describe any relationships you notice between each pair of problems and quotients.

   a. $32 \div 8 =$ _____ $3.2 \div 8 =$ _____

   b. $81 \div 9 =$ _____ $0.081 \div 9 =$ _____

4. Are the quotients below reasonable? Explain your answers.

   a. $5.6 \div 7 = 8$

   b. $56 \div 7 = 0.8$

   c. $0.56 \div 7 = 0.08$

Lesson 13:   Divide decimals by single-digit whole numbers involving easily
              identifiable multiples using place value understanding and relate
              to a written method.

EUREKA
MATH

5. 12.48 milliliters of medicine were separated into doses of 4 mL each. How many doses were made?

6. The price of milk in 2013 was around $3.28 a gallon. This was eight times as much as you would have probably paid for a gallon of milk in the 1950s. What was the cost for a gallon of milk during the 1950s? Use a tape diagram, and show your calculations.

Lesson 13:   Divide decimals by single-digit whole numbers involving easily
             identifiable multiples using place value understanding and relate
             to a written method.

63

©2015 Great Minds. eureka-math.org
G5-M1-SE-B1-1.3.1-01.2016

Name _____   Date _____

1. Complete the sentences with the correct number of units, and then complete the equation.

    a. 3 groups of _____ tenths is 1.5.   1.5 ÷ 3 = _____

    b. 6 groups of _____ hundredths is 0.24.   0.24 ÷ 6 = _____

    c. 5 groups of _____ thousandths is 0.045.   0.045 ÷ 5 = _____

2. Complete the number sentence. Express the quotient in units and then in standard form.

    a. 9.36 ÷ 3 = _____ ones ÷ 3 + _____ hundredths ÷ 3

        = _____ ones + _____ hundredths

        = _____

    b. 36.012 ÷ 3 = _____ ones ÷ 3 + _____ thousandths ÷ 3

        = _____ ones + _____ thousandths

        = _____

    c. 3.55 ÷ 5 = _____ tenths ÷ 5 + _____ hundredths ÷ 5

        = _____

        = _____

    d. 3.545 ÷ 5 = _____

        = _____

        = _____

Lesson 13:   Divide decimals by single-digit whole numbers involving easily
identifiable multiples using place value understanding and relate
to a written method.

EUREKA
MATH™

©2015 Great Minds. eureka-math.org
G5-M1-SE-B1-1.3.1-01.2016

3. Find the quotients. Then, use words, numbers, or pictures to describe any relationships you notice between each pair of problems and quotients.

   a.   21 ÷ 7 = _____   2.1 ÷ 7 = _____

   b.   48 ÷ 8 = _____   0.048 ÷ 8 = _____

4. Are the quotients below reasonable? Explain your answers.

   a.   0.54 ÷ 6 = 9

   b.   5.4 ÷ 6 = 0.9

   c.   54 ÷ 6 = 0.09

EUREKA MATH   **Lesson 13:**   Divide decimals by single-digit whole numbers involving easily   **65**
                                        identifiable multiples using place value understanding and relate
                                        to a written method.

©2015 Great Minds. eureka-math.org
G5-M1-SE-B1-1.3.1-01.2016

5. A toy airplane costs $4.84. It costs 4 times as much as a toy car. What is the cost of the toy car?

6. Julian bought 3.9 liters of cranberry juice, and Jay bought 8.74 liters of apple juice. They mixed the two juices together and then poured them equally into 2 bottles. How many liters of juice are in each bottle?

Lesson 13:   Divide decimals by single-digit whole numbers involving easily identifiable multiples using place value understanding and relate to a written method.

EUREKA
MATH™

Name _____   Date _____

1. Draw place value disks on the place value chart to solve. Show each step using the standard algorithm.

   a. $4.236 \div 3 =$ _____

Ones	Tenths	Hundredths	Thousandths

   $$3\overline{)4.236}$$

   b. $1.324 \div 2 =$ _____

Ones	Tenths	Hundredths	Thousandths

   $$2\overline{)1.324}$$

EUREKA MATH

Lesson 14:   Divide decimals with a remainder using place value understanding and relate to a written method.

67

2. Solve using the standard algorithm.

a. $0.78 \div 3 =$ _____	b. $7.28 \div 4 =$ _____	c. $17.45 \div 5 =$ _____

3. Grayson wrote $1.47 \div 7 = 2.1$ in her math journal.
   Use words, numbers, or pictures to explain why Grayson's thinking is incorrect.

4. Mrs. Nguyen used 1.48 meters of netting to make 4 identical mini hockey goals. How much netting did she use per goal?

5. Esperanza usually buys avocados for $0.94 apiece. During a sale, she gets 5 avocados for $4.10. How much money did she save per avocado? Use a tape diagram, and show your calculations.

EUREKA
MATH™

Name _____   Date _____

1. Draw place value disks on the place value chart to solve. Show each step using the standard algorithm.

    a. 5.241 ÷ 3 = _____

Ones	Tenths	Hundredths	Thousandths

$$3\overline{)5.241}$$

    b. 5.372 ÷ 4 = _____

Ones	Tenths	Hundredths	Thousandths

$$4\overline{)5.372}$$

EUREKA MATH

Lesson 14: Divide decimals with a remainder using place value understanding and relate to a written method.

69

2. Solve using the standard algorithm.

a. $0.64 \div 4 =$ _____	b. $6.45 \div 5 =$ _____	c. $16.404 \div 6 =$ _____

3. Mrs. Mayuko paid $40.68 for 3 kg of shrimp. What's the cost of 1 kilogram of shrimp?

4. The total weight of 6 pieces of butter and a bag of sugar is 3.8 lb. If the weight of the bag of sugar is 1.4 lb, what is the weight of each piece of butter?

Lesson 14: Divide decimals with a remainder using place value understanding and relate to a written method.

EUREKA MATH

Name _____ Date _____

1. Draw place value disks on the place value chart to solve. Show each step in the standard algorithm.

   a. $0.5 \div 2 =$ _____

Ones	•	Tenths	Hundredths	Thousandths

$$2\overline{)0.5}$$

   b. $5.7 \div 4 =$ _____

Ones	•	Tenths	Hundredths	Thousandths

$$4\overline{)5.7}$$

Lesson 15: Divide decimals using place value understanding, including remainders in the smallest unit.

71

EUREKA MATH

2. Solve using the standard algorithm.

a. $0.9 \div 2 =$	b. $9.1 \div 5 =$	c. $9 \div 6 =$
d. $0.98 \div 4 =$	e. $9.3 \div 6 =$	f. $91 \div 4 =$

3. Six bakers shared 7.5 kilograms of flour equally. How much flour did they each receive?

4. Mrs. Henderson makes punch by mixing 10.9 liters of apple juice, 0.6 liters of orange juice, and 8 liters of ginger ale. She pours the mixture equally into 6 large punch bowls. How much punch is in each bowl? Express your answer in liters.

Lesson 15:   Divide decimals using place value understanding, including remainders in the smallest unit.

EUREKA MATH

©2015 Great Minds. eureka-math.org
G5-M1-SE-B1-1.3.1-01.2016

Name _____   Date _____

1. Draw place value disks on the place value chart to solve. Show each step in the standard algorithm.

a.   0.7 ÷ 4 = _____

Ones	•	Tenths	Hundredths	Thousandths

$$4\,\overline{)\,0\,.\,7}$$

b.   8.1 ÷ 5 = _____

Ones	•	Tenths	Hundredths	Thousandths

$$5\,\overline{)\,8\,.\,1}$$

2. Solve using the standard algorithm.

a. 0.7 ÷ 2 =	b. 3.9 ÷ 6 =	c. 9 ÷ 4 =
d. 0.92 ÷ 2 =	e. 9.4 ÷ 4 =	f. 91 ÷ 8 =

3. A rope 8.7 meters long is cut into 5 equal pieces. How long is each piece?

4. Yasmine bought 6 gallons of apple juice. After filling up 4 bottles of the same size with apple juice, she had 0.3 gallon of apple juice left. How many gallons of apple juice are in each container?

Lesson 15:   Divide decimals using place value understanding, including remainders
                                  in the smallest unit.

EUREKA MATH

Name _____   Date _____

Solve.

1. Mr. Frye distributed $126 equally among his 4 children for their weekly allowance.

   a. How much money did each child receive?

   b. John, the oldest child, paid his siblings to do his chores. If John pays his allowance equally to his brother and two sisters, how much money will each of his siblings have received in all?

2. Ava is 23 cm taller than Olivia, and Olivia is half the height of Lucas. If Lucas is 1.78 m tall, how tall are Ava and Olivia? Express their heights in centimeters.

3. Mr. Hower can buy a computer with a down payment of $510 and 8 monthly payments of $35.75. If he pays cash for the computer, the cost is $699.99. How much money will he save if he pays cash for the computer instead of paying for it in monthly payments?

4. Brandon mixed 6.83 lb of cashews with 3.57 lb of pistachios. After filling up 6 bags that were the same size with the mixture, he had 0.35 lb of nuts left. What was the weight of each bag? Use a tape diagram, and show your calculations.

Lesson 16: Solve word problems using decimal operations.

5. The bakery bought 4 bags of flour containing 3.5 kg each. 0.475 kg of flour is needed to make a batch of muffins, and 0.65 kg is needed to make a loaf of bread.

a. If 4 batches of muffins and 5 loaves of bread are baked, how much flour will be left? Give your answer in kilograms.

b. The remaining flour is stored in bins that hold 3 kg each. How many bins will be needed to store the flour? Explain your answer.

Name _____ Date _____

Solve using tape diagrams.

1. A gardener installed 42.6 meters of fencing in a week. He installed 13.45 meters on Monday and 9.5 meters on Tuesday. He installed the rest of the fence in equal lengths on Wednesday through Friday. How many meters of fencing did he install on each of the last three days?

2. Jenny charges $9.15 an hour to babysit toddlers and $7.45 an hour to babysit school-aged children.

   a. If Jenny babysat toddlers for 9 hours and school-aged children for 6 hours, how much money did she earn in all?

   b. Jenny wants to earn $1,300 by the end of the summer. How much more will she need to earn to meet her goal?

Lesson 16: Solve word problems using decimal operations.

EUREKA MATH™

3. A table and 8 chairs weigh 235.68 lb together. If the table weighs 157.84 lb, what is the weight of one chair in pounds?

4. Mrs. Cleaver mixes 1.24 liters of red paint with 3 times as much blue paint to make purple paint. She pours the paint equally into 5 containers. How much blue paint is in each container? Give your answer in liters.

This page intentionally left blank

Eureka Math
Grade 5
Module 2

Special thanks go to the Gordon A. Cain Center and to the Department of Mathematics at Louisiana State University for their support in the development of Eureka Math.

Name _____   Date _____

1. Fill in the blanks using your knowledge of place value units and basic facts.

a. 23 × 20 Think: 23 ones × 2 tens = _____ tens 23 × 20 = _____	b. 230 × 20 Think: 23 tens × 2 tens = _____ 230 × 20 = _____
c. 41 × 4 41 ones × 4 ones = 164 _____ 41 × 4 = _____	d. 410 × 400 41 tens × 4 hundreds = 164 _____ 410 × 400 = _____
e. 3,310 × 300 _____ tens × _____ hundreds = 993 _____ 3,310 × 300 = _____	f. 500 × 600 _____ hundreds × _____ hundreds = 3? _____ 500 × 600 = _____

2. Determine if these equations are true or false. Defend your answer using your knowledge of place value and the commutative, associative, and/or distributive properties.

   a. 6 tens = 2 tens × 3 tens

   b. 44 × 20 × 10 = 440 × 2

   c. 86 ones × 90 hundreds = 86 ones × 900 tens

   d. 64 × 8 × 100 = 640 × 8 × 10

EUREKA MATH™

Lesson 1:   Multiply multi-digit whole numbers and multiples of 10 using place value patterns and the distributive and associative properties.

1

e. $57 \times 2 \times 10 \times 10 \times 10 = 570 \times 2 \times 10$

3. Find the products. Show your thinking. The first row gives some ideas for showing your thinking.

a.
7×9	7×90	70×90	70×900
$= 63$	$= 63 \times 10$	$= (7 \times 10) \times (9 \times 10)$	$= (7 \times 9) \times (10 \times 100)$
	$= 630$	$= (7 \times 9) \times 100$	$= 63,000$
		$= 6,300$	

b.
45×3	45×30	450×30	450×300

c.
40×5	40×50	40×500	$400 \times 5,000$

d.
718×2	$7,180 \times 20$	$7,180 \times 200$	$71,800 \times 2,000$

Lesson 1: Multiply multi-digit whole numbers and multiples of 10 using place value patterns and the distributive and associative properties.

EUREKA MATH

4. Ripley told his mom that multiplying whole numbers by multiples of 10 was easy because you just count zeros in the factors and put them in the product. He used these two examples to explain his strategy.

$$7,000 \times 600 = 4,200,000 \qquad\qquad 800 \times 700 = 560,000$$
(3 zeros) (2 zeros) (5 zeros) (2 zeros) (2 zeros) (4 zeros)

Ripley's mom said his strategy will not always work. Why not? Give an example.

5. The Canadian side of Niagara Falls has a flow rate of 600,000 gallons per second. How many gallons of water flow over the falls in 1 minute?

6. Tickets to a baseball game are $20 for an adult and $15 for a student. A school buys tickets for 45 adults and 600 students. How much money will the school spend for the tickets?

Lesson 1: Multiply multi-digit whole numbers and multiples of 10 using place value patterns and the distributive and associative properties.

3

Name _____ Date _____

1. Fill in the blanks using your knowledge of place value units and basic facts.

   a. 43×30

      Think: 43 ones × 3 tens = _____ tens

      $43 \times 30 =$ _____

   b. 430×30

      Think: 43 tens × 3 tens = _____ hundreds

      $430 \times 30 =$ _____

   c. 830×20

      Think: 83 tens × 2 tens = 166 _____

      $830 \times 20 =$ _____

   d. $4{,}400 \times 400$

      _____ hundreds × _____ hundreds = 176 _____

      $4{,}400 \times 400 =$ _____

   e. $80 \times 5{,}000$

      _____ tens × _____ thousands = 40 _____

      $80 \times 5{,}000 =$ _____

2. Determine if these equations are true or false. Defend your answer using your knowledge of place value and the commutative, associative, and/or distributive properties.

   a. 35 hundreds = 5 tens × 7 tens

   b. $770 \times 6 = 77 \times 6 \times 100$

   c. 50 tens × 4 hundreds = 40 tens × 5 hundreds

   d. $24 \times 10 \times 90 = 90 \times 2{,}400$

Lesson 1: Multiply multi-digit whole numbers and multiples of 10 using place value patterns and the distributive and associative properties.

EUREKA MATH™

3. Find the products. Show your thinking. The first row gives some ideas for showing your thinking.

   a. 5×5 5×50 50×50 50×500

       $= 25$ $= 25 \times 10$ $= (5 \times 10) \times (5 \times 10)$ $= (5 \times 5) \times (10 \times 100)$

                            $= 250$ $= (5 \times 5) \times 100$ $= 25,000$

                                           $= 2,500$

   b. 80×5 80×50 800×500 $8,000 \times 50$

   c. 637×3 $6,370 \times 30$ $6,370 \times 300$ $63,700 \times 300$

4. A concrete stepping-stone measures 20 square inches. What is the area of 30 such stones?

5. A number is 42,300 when multiplied by 10. Find the product of this number and 500.

Lesson 1:   Multiply multi-digit whole numbers and multiples of 10 using place value patterns and the distributive and associative properties.

5

This page intentionally left blank

$\frac{1}{1,000}$	Thousandths					
$\frac{1}{100}$	Hundredths					
$\frac{1}{10}$	Tenths					
•	•	•	•	•	•	•	•
1	Ones					
10	Tens					
100	Hundreds					
1,000	Thousands					
10,000	Ten Thousands					
100,000	Hundred Thousands					
1,000,000	Millions					

millions to thousandths place value chart

EUREKA MATH™

Lesson 1: Multiply multi-digit whole numbers and multiples of 10 using place value patterns and the distributive and associative properties.

7

This page intentionally left blank

Name _____ Date _____

1. Round the factors to estimate the products.

    a. ⁻97 × 52 ≈ _____ × _____ = _____

    A reasonable estimate ᵢ. ⁵97 × 52 is _____.

    b. 1,103 × 59 ≈ _____ × _____ = _____

    A reasonable estimate for 1,103 × 59 is _____.

    c. 5,840 × 25 ≈ _____ × _____ = _____

    A reasonable estimate for 5,840 × 25 is _____.

2. Complete the table using your understanding of place value and knowledge of rounding to estimate the product.

	Expressions	Rounded Factors	Estimate
a.	2,809 × 42	3,000 × 40	120,000
b.	28,090 × 420		
c.	8,932 × 59		
d.	89 tens × 63 tens		
e.	398 hundreds × 52 tens		

EUREKA MATH™

Lesson 2: Estimate multi-digit products by rounding factors to a basic fact and using place value patterns.

9

©2015 Great Minds. eureka-math.org
G5-M2-SE-B1-1.3.1-01.2016

3. For which of the following expressions would 200,000 be a reasonable estimate? Explain how you know.

    2,146 × 12   21,467 × 121   2,146 × 121   21,477 × 1,217

4. Fill in the missing factors to find the given estimated product.

    a. 571 × 43 ≈ _____ × _____ = 24,000

    b. 726 × 674 ≈ _____ × _____ = 490,000

    c. 8,379 × 541 ≈ _____ × _____ = 4,000,000

5. There are 19,763 tickets available for a New York Knicks home game. If there are 41 home games in a season, about how many tickets are available for all the Knicks' home games?

6. Michael saves $423 dollars a month for college.

    a. About how much money will he have saved after 4 years?

    b. Will your estimate be lower or higher than the actual amount Michael will save? How do you know?

Lesson 2:   Estimate multi-digit products by rounding factors to a basic fact and
              using place value patterns.

EUREKA
MATH™

Name _____   Date _____

1. Round the factors to estimate the products.

   a. $697 \times 82 \approx$ _____ $\times$ _____ $=$ _____

      A reasonable estimate for 697×82 is _____.

   b. $5,897 \times 67 \approx$ _____ $\times$ _____ $=$ _____

      A reasonable estimate for $5,897 \times 67$ is _____.

   c. $8,840 \times 45 \approx$ _____ $\times$ _____ $=$ _____

      A reasonable estimate for $8,840 \times 45$ is _____.

2. Complete the table using your understanding of place value and knowledge of rounding to estimate the product.

Expressions	Rounded Factors	Estimate
a. $3,409 \times 73$	$3,000 \times 70$	$210,000$
b. $82,290 \times 240$		
c. $9,832 \times 39$		
d. 98 tens $\times$ 36 tens		
e. 893 hundreds $\times$ 85 tens		

3. The estimated answer to a multiplication problem is 800,000. Which of the following expressions could result in this answer? Explain how you know.

       $8,146 \times 12$ $81,467 \times 121$ $8,146 \times 121$ $81,477 \times 1,217$

4. Fill in the blank with the missing estimate.

   a. 751 × 34 ≈ _____ × _____ = 24,000

   b. 627 × 674 ≈ _____ × _____ = 420,000

   c. 7,939 × 541 ≈ _____ × _____ = 4,000,000

5. In a single season, the New York Yankees sell an average of 42,362 tickets for each of their 81 home games. About how many tickets do they sell for an entire season of home games?

6. Raphael wants to buy a new car.

   a. He needs a down payment of $3,000. If he saves $340 each month, about how many months will it take him to save the down payment?

   b. His new car payment will be $288 each month for five years. What is the total of these payments?

Lesson 2:   Estimate multi-digit products by rounding factors to a basic fact and using place value patterns.

EUREKA MATH™

Name _____ Date _____

1. Draw a model. Then, write the numerical expressions.

a. The sum of 8 and 7, doubled	b. 4 times the sum of 14 and 26
c. 3 times the difference between 37.5 and 24.5	d. The sum of 3 sixteens and 2 nines
e. The difference between 4 twenty-fives and 3 twenty-fives	f. Triple the sum of 33 and 27

 EUREKA MATH™

Lesson 3:   Write and interpret numerical expressions, and compare expressions
using a visual model.

13

©2015 Great Minds. eureka-math.org
G5-M2-SE-B1-1.3.1-01.2016

2. Write the numerical expressions in words. Then, solve.

Expression	Words	The Value of the Expression
a. $12 \times (5 + 25)$		
b. $(62 - 12) \times 11$		
c. $(45 + 55) \times 23$		
d. $(30 \times 2) + (8 \times 2)$		

3. Compare the two expressions using > , < , or = . In the space beneath each pair of expressions, explain how you can compare without calculating. Draw a model if it helps you.

a. $24 \times (20 + 5)$	◯	$(20 + 5) \times 12$
b. 18×27	◯	20 twenty-sevens minus 1 twenty-seven
c. 19×9	◯	3 nineteens, tripled

Lesson 3:   Write and interpret numerical expressions, and compare expressions using a visual model.

EUREKA MATH™

4. Mr. Huynh wrote *the sum of 7 fifteens and 38 fifteens* on the board.

   Draw a model, and write the correct expression.

5. Two students wrote the following numerical expressions.

   Angeline: (7 + 15) × (38 + 15)

   MeiLing: 15 × (7 + 38)

   Are the students' expressions equivalent to your answer in Problem 4? Explain your answer.

6. A box contains 24 oranges. Mr. Lee ordered 8 boxes for his store and 12 boxes for his restaurant.
   a. Write an expression to show how to find the total number of oranges ordered.

   b. Next week, Mr. Lee will double the number of boxes he orders. Write a new expression to represent the number of oranges in next week's order.

   c. Evaluate your expression from Part (b) to find the total number of oranges ordered in both weeks.

EUREKA MATH   Lesson 3:   Write and interpret numerical expressions, and compare expressions   15
                                      using a visual model.

©2015 Great Minds. eureka-math.org
G5-M2-SE-B1-1.3.1-01.2016

Name _____ Date _____

1. Draw a model. Then, write the numerical expressions.

a. The sum of 21 and 4, doubled	b. 5 times the sum of 7 and 23
c. 2 times the difference between 49.5 and 37.5	d. The sum of 3 fifteens and 4 twos
e. The difference between 9 thirty-sevens and 8 thirty-sevens	f. Triple the sum of 45 and 55

Lesson 3:   Write and interpret numerical expressions, and compare expressions using a visual model.

EUREKA MATH™

©2015 Great Minds. eureka-math.org
G5-M2-SE-B1-1.3.1-01.2016

2. Write the numerical expressions in words. Then, solve.

Expression	Words	The Value of the Expression
a. $10 \times (2.5 + 13.5)$		
b. $(98 - 78) \times 11$		
c. $(71 + 29) \times 26$		
d. $(50 \times 2) + (15 \times 2)$		

3. Compare the two expressions using > , < , or = . In the space beneath each pair of expressions, explain how you can compare without calculating. Draw a model if it helps you.

a. $93 \times (40 + 2)$	$\bigcirc$	$(40 + 2) \times 39$
b. 61×25	$\bigcirc$	60 twenty-fives minus 1 twenty-five

4. Larry claims that (14 + 12) × (8 + 12) and (14 × 12) + (8 × 12) are equivalent because they have the same digits and the same operations.

   a. Is Larry correct? Explain your thinking.

   b. Which expression is greater? How much greater?

Lesson 3: Write and interpret numerical expressions, and compare expressions using a visual model.

EUREKA MATH

Name _____   Date _____

1. Circle each expression that is not equivalent to the expression in **bold**.

    a. **16 × 29**

        29 sixteens   16 × (30 − 1)   (15 − 1) × 29   (10 × 29) − (6 × 29)

    b. **38 × 45**

        (38 + 40) × (38 + 5)   (38 × 40) + (38 × 5)   45 × (40 + 2)   45 thirty-eights

    c. **74 × 59**

        74 × (50 + 9)   74 × (60 − 1)   (74 × 5) + (74 × 9)   59 seventy-fours

2. Solve using mental math. Draw a tape diagram and fill in the blanks to show your thinking. The first one is partially done for you.

    a. 19 × 25 = _____ twenty-fives

25	25	25	...	25	25
1	2	3	...	19	20

        Think: 20 twenty-fives − 1 twenty-five.

        = (_____ × 25) − (_____ × 25)

        = _____ − _____

        = _____

    b. 24 × 11 = _____ twenty-fours

        Think: _____ twenty fours + _____ twenty four

        = (_____ × 24) + (_____ × 24)

        = _____ + _____

        = _____

c. $79 \times 14 =$ _____ fourteens

Think: _____ fourteens − 1 fourteen

$= ($_____ $\times 14) - ($_____ $\times 14)$

$=$ _____ $-$ _____

$=$ _____

d. $21 \times 75 =$ _____ seventy-fives

Think: _____ seventy-fives + _____ seventy-five

$= ($_____ $\times 75) + ($_____ $\times 75)$

$=$ _____ $+$ _____

$=$ _____

3. Define the unit in word form and complete the sequence of problems as was done in the lesson.

a. $19 \times 15 = 19$ _____

Think: 20 _____ $- 1$ _____

$= (20 \times$ _____ $) - (1 \times$ _____ $)$

$=$ _____ $-$ _____

$=$ _____

b. $14 \times 15 = 14$ _____

Think: 10 _____ $+ 4$ _____

$= (10 \times$ _____ $) + (4 \times$ _____ $)$

$=$ _____ $+$ _____

$=$ _____

Lesson 4:   Convert numerical expressions into unit form as a mental strategy for multi-digit multiplication.

EUREKA MATH™

c. $25 \times 12 = 12$ _____

Think: 10 _____ + 2 _____

= (10 × _____) + (2 × _____)

= _____ + _____

= _____

d. $18 \times 17 = 18$ _____

Think: 20 _____ − 2 _____

= (20 × _____) − (2 × _____)

= _____ − _____

= _____

4. How can 14×50 help you find 14×49?

5. Solve mentally.

   a. $101 \times 15 =$ _____

   b. $18 \times 99 =$ _____

6. Saleem says 45×32 is the same as $(45 \times 3) + (45 \times 2)$. Explain Saleem's error using words, numbers, and/or pictures.

7. Juan delivers 174 newspapers every day. Edward delivers 126 more newspapers each day than Juan.

   a. Write an expression to show how many newspapers Edward will deliver in 29 days.

   b. Use mental math to solve. Show your thinking.

EUREKA MATH

Lesson 4: Convert numerical expressions into unit form as a mental strategy for multi-digit multiplication.

21

©2015 Great Minds. eureka-math.org
G5-M2-SE-B1-1.3.1-01.2016

Name _____ Date _____

1. Circle each expression that is not equivalent to the expression in **bold**.

   a. **37 × 19**

   37 nineteens   (30 × 19) − (7 × 29)   37 × (20 − 1)   (40 − 2) × 19

   b. **26 × 35**

   35 twenty-sixes   (26 + 30) × (26 + 5)   (26 × 30) + (26 × 5)   35 × (20 + 60)

   c. **34 × 89**

   34 × (80 + 9)   (34 × 8) + (34 × 9)   34 × (90 − 1)   89 thirty-fours

2. Solve using mental math. Draw a tape diagram and fill in the blanks to show your thinking. The first one is partially done for you.

   a. 19 × 50 = _____ fifties

50	50	50	...	50	⊠
1	2	3	...	19	20

   Think: 20 fifties − 1 fifty

   = (_____ × 50) − (_____ × 50)

   = _____ − _____

   = _____

   b. 11 × 26 = _____ twenty-sixes

   Think: _____ twenty-sixes + _____ twenty-six

   = (_____ × 26) + (_____ × 26)

   = _____ + _____

   = _____

Lesson 4:   Convert numerical expressions into unit form as a mental strategy for multi-digit multiplication.

EUREKA MATH

c.   49 × 12 = _____ twelves

d.   12 × 25 = _____ twenty-fives

Think: _____ twelves − 1 twelve

   = (_____ × 12) − (_____ × 12)

   = _____ − _____

   = _____

Think: _____ twenty-fives + _____ twenty-fives

   = (_____ × 25) + (_____ × 25)

   = _____ + _____

   = _____

3.   Define the unit in word form and complete the sequence of problems as was done in the lesson.

a.   29 × 12 = 29 _____

b.   11 × 31 = 31 _____

Think: 30 _____ − 1 _____

   = (30 × _____) − (1 × _____)

   = _____ − _____

   = _____

Think: 30 _____ + 1 _____

   = (30 × _____) + (1 × _____)

   = _____ + _____

   = _____

EUREKA
MATH™

Lesson 4:   Convert numerical expressions into unit form as a mental strategy
for multi-digit multiplication.

23

©2015 Great Minds. eureka-math.org
G5-M2-SE-B1-1.3.1-01.2016

c. $19 \times 11 = 19$ _____

d. $50 \times 13 = 13$ _____

Think: 20 _____ $- 1$ _____

     $= (20 \times$ _____ $) - (1 \times$ _____ $)$

     $=$ _____ $-$ _____

     $=$ _____

Think: 10 _____ $+ 3$ _____

     $= (10 \times$ _____ $) + (3 \times$ _____ $)$

     $=$ _____ $+$ _____

     $=$ _____

4.   How can 12×50 help you find 12×49?

5.   Solve mentally.

    a. $16 \times 99 =$ _____

    b. $20 \times 101 =$ _____

6.   Joy is helping her father to build a rectangular deck that measures 14 ft by 19 ft. Find the area of the deck using a mental strategy. Explain your thinking.

7.   The Lason School turns 101 years old in June. In order to celebrate, they ask each of the 23 classes to collect 101 items and make a collage. How many total items will be in the collage? Use mental math to solve. Explain your thinking.

Lesson 4:   Convert numerical expressions into unit form as a mental strategy for multi-digit multiplication.

EUREKA MATH™

Name _____   Date _____

1. Draw an area model, and then solve using the standard algorithm. Use arrows to match the partial products from the area model to the partial products of the algorithm.

a. 34 × 21 = _____

$$
\begin{array}{r}
3\,4 \\
\times\ 2\,1 \\
\hline
\end{array}
$$

b. 434 × 21 = _____

$$
\begin{array}{r}
4\,3\,4 \\
\times\ \ 2\,1 \\
\hline
\end{array}
$$

2. Solve using the standard algorithm.

a. 431 × 12 = _____   b. 123 × 23 = _____   c. 312 × 32 = _____

EUREKA MATH™   Lesson 5:   Connect visual models and the distributive property to partial products
                     of the standard algorithm without renaming.   25

©2015 Great Minds. eureka-math.org
G5-M2-SE-B1-1.3.1-01.2016

3. Betty saves $161 a month. She saves $141 less each month than Jack. How much will Jack save in 2 years?

4. Farmer Brown feeds 12.1 kilograms of alfalfa to each of his 2 horses daily. How many kilograms of alfalfa will all his horses have eaten after 21 days? Draw an area model to solve.

Lesson 5: Connect visual models and the distributive property to partial products of the standard algorithm without renaming.

©2015 Great Minds. eureka-math.org
G5-M2-SE-B1-1.3.1-01.2016

Name _____ Date _____

1. Draw an area model, and then solve using the standard algorithm. Use arrows to match the partial products from the area model to the partial products in the algorithm.

   a. $24 \times 21 =$ 44

   $$\begin{array}{r} 2\,4 \\ \times\ 2\,1 \\ \hline \end{array}$$

   b. $242 \times 21 =$ 282

   $$\begin{array}{r} 2\,4\,2 \\ \times\ \ 2\,1 \\ \hline \end{array}$$

2. Solve using the standard algorithm.

   a. $314 \times 22 =$ 318

   b. $413 \times 22 =$ _____

   c. $213 \times 32 =$ _____

Lesson 5: Connect visual models and the distributive property to partial products of the standard algorithm without renaming.

27

EUREKA MATH™

3. A young snake measures 0.23 meters long. During the course of his lifetime, he will grow to be 13 times his current length. What will his length be when he is full grown?

4. Zenin earns $142 per shift at his new job. During a pay period, he works 12 shifts. What would his pay be for that period?

Lesson 5: Connect visual models and the distributive property to partial products of the standard algorithm without renaming.

EUREKA MATH

Name _____   Date _____

1. Draw an area model. Then, solve using the standard algorithm. Use arrows to match the partial products from your area model to the partial products in the algorithm.

a. 48 × 35

$$
\begin{array}{r}
4\,8 \\
\times\ 3\,5 \\
\hline
\end{array}
$$

b. 648 × 35

$$
\begin{array}{r}
6\,4\,8 \\
\times\ 3\,5 \\
\hline
\end{array}
$$

EUREKA
MATH™

Lesson 6:   Connect area models and the distributive property to partial products
            of the standard algorithm with renaming.

29

2. Solve using the standard algorithm.

   a. 758×92 b. 958×94

   c. 476×65 d. 547×64

3. Carpet costs \$16 a square foot. A rectangular floor is 16 feet long by 14 feet wide. How much would it cost to carpet the floor?

Lesson 6:   Connect area models and the distributive property to partial products of the standard algorithm with renaming.

EUREKA MATH™

4. General admission to The American Museum of Natural History is $19.

   a. If a group of 125 students visits the museum, how much will the group's tickets cost?

   b. If the group also purchases IMAX movie tickets for an additional $4 per student, what is the new total cost of all the tickets? Write an expression that shows how you calculated the new price.

     Lesson 6:   Connect area models and the distributive property to partial products   31
                          of the standard algorithm with renaming.

©2015 Great Minds. eureka-math.org
G5-M2-SE-B1-1.3.1-01.2016

Name _____   Date _____

1. Draw an area model. Then, solve using the standard algorithm. Use arrows to match the partial products from your area model to the partial products in the algorithm.

   a. 27 × 36

$$\begin{array}{r} 27 \\ \times\ 36 \\ \hline \end{array}$$

   b. 527 × 36

$$\begin{array}{r} 527 \\ \times\ 36 \\ \hline \end{array}$$

Lesson 6:   Connect area models and the distributive property to partial products of the standard algorithm with renaming.

EUREKA MATH

2. Solve using the standard algorithm.

   a. 649×53

   b. 496×53

   c. 758×46

   d. 529×48

3. Each of the 25 students in Mr. McDonald's class sold 16 raffle tickets. If each ticket costs $15, how much money did Mr. McDonald's students raise?

EUREKA MATH

Lesson 6: Connect area models and the distributive property to partial products
           of the standard algorithm with renaming.

33

4. Jayson buys a car and pays by installments. Each installment is $567 per month. After 48 months, Jayson owes $1,250. What was the total price of the vehicle?

EUREKA MATH™

Name _____   Date _____

1. Draw an area model. Then, solve using the standard algorithm. Use arrows to match the partial products from the area model to the partial products in the algorithm.

   a.   481 × 352

$$\begin{array}{r} 481 \\ \times\ 352 \\ \hline \end{array}$$

   b.   481 × 302

$$\begin{array}{r} 481 \\ \times\ 302 \\ \hline \end{array}$$

   c.   Why are there three partial products in 1(a) and only two partial products in 1(b)?

©2015 Great Minds. eureka-math.org
G5-M2-SE-B1-1.3.1-01.2016

2. Solve by drawing the area model and using the standard algorithm.

   a. 8,401 × 305

$$\begin{array}{r} 8,401 \\ \times\ \ \ 305 \\ \hline \end{array}$$

   b. 7,481 × 350

$$\begin{array}{r} 7,481 \\ \times\ \ \ 350 \\ \hline \end{array}$$

3. Solve using the standard algorithm.

   a. 346 × 27

   b. 1,346 × 297

EUREKA
MATH™

c.   346 × 207

d.   1,346 × 207

4. A school district purchased 615 new laptops for their mobile labs. Each computer cost $409. What is the total cost for all of the laptops?

5. A publisher prints 1,512 copies of a book in each print run. If they print 305 runs, how many books will be printed?

6. As of the 2010 census, there were 3,669 people living in Marlboro, New York. Brooklyn, New York, has 681 times as many people. How many more people live in Brooklyn than in Marlboro?

Lesson 7:   Connect area models and the distributive property to partial products of the standard algorithm with renaming.

37

©2015 Great Minds. eureka-math.org
G5-M2-SE-B1-1.3.1-01.2016

Name _____   Date _____

1. Draw an area model. Then, solve using the standard algorithm. Use arrows to match the partial products from your area model to the partial products in your algorithm.

   a. 273×346

$$\begin{array}{r} 273 \\ \times\ 346 \\ \hline \end{array}$$

   b. 273×306

$$\begin{array}{r} 273 \\ \times\ 306 \\ \hline \end{array}$$

   c. Both Parts (a) and (b) have three-digit multipliers. Why are there three partial products in Part (a) and only two partial products in Part (b)?

Lesson 7:   Connect area models and the distributive property to partial products of the standard algorithm with renaming.

EUREKA MATH™

2. Solve by drawing the area model and using the standard algorithm.

   a. 7,481 × 290

   b. 7,018 × 209

3. Solve using the standard algorithm.
   a. 426 × 357

   b. 1,426 × 357

Lesson 7:   Connect area models and the distributive property to partial products
             of the standard algorithm with renaming.

39

©2015 Great Minds. eureka-math.org
G5-M2-SE-B1-1.3.1-01.2016

c. 426 × 307

d. 1,426 × 307

4. The Hudson Valley Renegades Stadium holds a maximum of 4,505 people. During the height of their popularity, they sold out 219 consecutive games. How many tickets were sold during this time?

5. One Saturday at the farmer's market, each of the 94 vendors made $502 in profit. How much profit did all vendors make that Saturday?

Lesson 7:   Connect area models and the distributive property to partial products of the standard algorithm with renaming.

©2015 Great Minds. eureka-math.org
G5-M2-SE-B1-1.3.1-01.2016

Name _____   Date _____

1. Estimate the product first. Solve by using the standard algorithm. Use your estimate to check the reasonableness of the product.

a. 213 × 328 ≈ 200 × 300 = 60,000 　 2 1 3 × 　3 2 8	b. 662 × 372	c. 739 × 442
d. 807 × 491	e. 3,502 × 656	f. 4,390 × 741
g. 530 × 2,075	h. 4,004 × 603	i. 987 × 3,105

Lesson 8:   Fluently multiply multi-digit whole numbers using the standard algorithm and using estimation to check for reasonableness of the product.

41

©2015 Great Minds. eureka-math.org
G5-M2-SE-B1-1.3.1-01.2016

2. Each container holds 1 L 275 mL of water. How much water is in 609 identical containers? Find the difference between your estimated product and precise product.

3. A club had some money to purchase new chairs. After buying 355 chairs at $199 each, there was $1,068 remaining. How much money did the club have at first?

Lesson 8:   Fluently multiply multi-digit whole numbers using the standard algorithm and using estimation to check for reasonableness of the product.

©2015 Great Minds. eureka-math.org
G5-M2-SE-B1-1.3.1-01.2016

4. So far, Carmella has collected 14 boxes of baseball cards. There are 315 cards in each box. Carmella estimates that she has about 3,000 cards, so she buys 6 albums that hold 500 cards each.

   a. Will the albums have enough space for all of her cards? Why or why not?

   b. How many cards does Carmella have?

   c. How many albums will she need for all of her baseball cards?

Lesson 8: Fluently multiply multi-digit whole numbers using the standard algorithm and using estimation to check for reasonableness of the product.

43

Name _____ Date _____

1. Estimate the product first. Solve by using the standard algorithm. Use your estimate to check the reasonableness of the product.

a. 312 × 149 ≈ 300 × 100 = 30,000 3 1 2 × 1 4 9	b. 743 × 295	c. 428 × 637
d. 691 × 305	e. 4,208 × 606	f. 3,068 × 523
g. 430 × 3,064	h. 3,007 × 502	i. 254 × 6,104

Lesson 8:   Fluently multiply multi-digit whole numbers using the standard algorithm and using estimation to check for reasonableness of the product.

2. When multiplying 1,729 times 308, Clayton got a product of 53,253. Without calculating, does his product seem reasonable? Explain your thinking.

3. A publisher prints 1,912 copies of a book in each print run. If they print 305 runs, the manager wants to know about how many books will be printed. What is a reasonable estimate?

EUREKA
MATH™

Lesson 8:   Fluently multiply multi-digit whole numbers using the standard algorithm and using estimation to check for reasonableness of the product.

45

©2015 Great Minds. eureka-math.org
G5-M2-SE-B1-1.3.1-01.2016

This page intentionally left blank

Name _____ Date _____

Solve.

1. An office space in New York City measures 48 feet by 56 feet. If it sells for $565 per square foot, what is the total cost of the office space?

2. Gemma and Leah are both jewelry makers. Gemma made 106 beaded necklaces. Leah made 39 more necklaces than Gemma.

   a. Each necklace they make has exactly 104 beads on it. How many beads did both girls use altogether while making their necklaces?

   b. At a recent craft fair, Gemma sold each of her necklaces for $14. Leah sold each of her necklaces for $10 more. Who made more money at the craft fair? How much more?

Lesson 9:   Fluently multiply multi-digit whole numbers using the standard
            algorithm to solve multi step word problems.

47

3. Peng bought 26 treadmills for her new fitness center at $1,334 each. Then, she bought 19 stationary bikes for $749 each. How much did she spend on her new equipment? Write an expression, and then solve.

4. A Hudson Valley farmer has 26 employees. He pays each employee $410 per week. After paying his workers for one week, the farmer has $162 left in his bank account. How much money did he have at first?

5. Frances is sewing a border around 2 rectangular tablecloths that each measure 9 feet long by 6 feet wide. If it takes her 3 minutes to sew on 1 inch of border, how many minutes will it take her to complete her sewing project? Write an expression, and then solve.

Lesson 9: Fluently multiply multi-digit whole numbers using the standard algorithm to solve multi step word problems.

©2015 Great Minds. eureka-math.org
G5-M2-SE-B1-1.3.1-01.2016

6. Each grade level at Hooperville Schools has 298 students.

    a. If there are 13 grade levels, how many students attend Hooperville Schools?

    b. A nearby district, Willington, is much larger. They have 12 times as many students. How many students attend schools in Willington?

Lesson 9:   Fluently multiply multi-digit whole numbers using the standard
              algorithm to solve multi step word problems.   49

©2015 Great Minds. eureka-math.org
G5-M2-SE-B1-1.3.1-01.2016

Name _____   Date _____

Solve.

1. Jeffery bought 203 sheets of stickers. Each sheet has a dozen stickers. He gave away 907 stickers to his family and friends on Valentine's Day. How many stickers does Jeffery have remaining?

2. During the 2011 season, a quarterback passed for 302 yards per game. He played in all 16 regular season games that year.

    a. For how many total yards did the quarterback pass?

    b. If he matches this passing total for each of the next 13 seasons, how many yards will he pass for in his career?

Lesson 9:   Fluently multiply multi-digit whole numbers using the standard algorithm to solve multi step word problems.

3. Bao saved $179 a month. He saved $145 less than Ada each month. How much would Ada save in three and a half years?

4. Mrs. Williams is knitting a blanket for her newborn granddaughter. The blanket is 2.25 meters long and 1.8 meters wide. What is the area of the blanket? Write the answer in centimeters.

Lesson 9: Fluently multiply multi-digit whole numbers using the standard algorithm to solve multi step word problems.

51

©2015 Great Minds. eureka-math.org
G5-M2-SE-B1-1.3.1-01.2016

5. Use the chart to solve.   **Soccer Field Dimensions**

	FIFA Regulation (in yards)	New York State High Schools (in yards)
Minimum Length	110	100
Maximum Length	120	120
Minimum Width	70	55
Maximum Width	80	80

a. Write an expression to find the difference in the maximum area and minimum area of a NYS high school soccer field. Then, evaluate your expression.

b. Would a field with a width of 75 yards and an area of 7,500 square yards be within FIFA regulation? Why or why not?

c. It costs $26 to fertilize, water, mow, and maintain each square yard of a full size FIFA field (with maximum dimensions) before each game. How much will it cost to prepare the field for next week's match?

Lesson 9:   Fluently multiply multi-digit whole numbers using the standard algorithm to solve multi step word problems.

EUREKA MATH

Name _____   Date _____

1. Estimate the product. Solve using an area model and the standard algorithm. Remember to express your products in standard form.

    a. 22 × 2.4 ≈ _____ × _____ = _____

    2 4 (tenths)

    × 2 2

    b. 3.1 × 33 _____ × _____ = _____

    3 1 (tenths)

    × 3 3

2. Estimate. Then, use the standard algorithm to solve. Express your products in standard form.

    a. 3.2 × 47 ≈ _____ × _____ = _____

    3 2 (tenths)

    × 4 7

    b. 3.2 × 94 ≈ _____ × _____ = _____

    3 2 (tenths)

    × 9 4

EUREKA MATH™

Lesson 10:   Multiply decimal fractions with tenths by multi-digit whole numbers
using place value understanding to record partial products.

53

©2015 Great Minds. eureka-math.org
G5-M2-SE-B1-1.3.1-01.2016

c. $6.3 \times 44 \approx$ _____ × _____ = _____

d. $14.6 \times 17 \approx$ _____ × _____ = _____

e. $8.2 \times 34 \approx$ _____ × _____ = _____

f. $160.4 \times 17 \approx$ _____ × _____ = _____

3.   Michelle multiplied 3.4×52. She incorrectly wrote 1,768 as her product. Use words, numbers, and/or pictures to explain Michelle's mistake.

4.   A wire is bent to form a square with a perimeter of 16.4 cm. How much wire would be needed to form 25 such squares? Express your answer in meters.

Lesson 10:   Multiply decimal fractions with tenths by multi-digit whole numbers using place value understanding to record partial products.

EUREKA MATH™

©2015 Great Minds. eureka-math.org
G5-M2-SE-B1-1.3.1-01.2016

Name _____   Date _____

1. Estimate the product. Solve using an area model and the standard algorithm. Remember to express your products in standard form.

   a. $53 \times 1.2 \approx$ _____ $\times$ _____ $=$ _____

                                                     1 2 (tenths)

                                                 $\times\,5\,3$

   b. $2.1 \times 82 \approx$ _____ $\times$ _____ $=$ _____

                                                     2 1 (tenths)

                                                 $\times\,8\,2$

2. Estimate. Then, use the standard algorithm to solve. Express your products in standard form.

   a. $4.2 \times 34 \approx$ _____ $\times$ _____ $=$ _____   b. $65 \times 5.8 \approx$ _____ $\times$ _____ $=$ _____

                     4 2 (tenths)   5 8 (tenths)

                       $\times\,3\,4$ $\times\,6\,5$

c. $3.3 \times 16 \approx$ _____ × _____ = _____

d. $15.6 \times 17 \approx$ _____ × _____ = _____

e. $73 \times 2.4 \approx$ _____ × _____ = _____

f. $193.5 \times 57 \approx$ _____ × _____ = _____

3. Mr. Jansen is building an ice rink in his backyard that will measure 8.4 meters by 22 meters. What is the area of the rink?

4. Rachel runs 3.2 miles each weekday and 1.5 miles each day of the weekend. How many miles will she have run in 6 weeks?

Lesson 10: Multiply decimal fractions with tenths by multi-digit whole numbers using place value understanding to record partial products.

EUREKA MATH

©2015 Great Minds. eureka-math.org
G5-M2-SE-B1-1.3.1-01.2016

Name _____   Date _____

1. Estimate the product. Solve using the standard algorithm. Use the thought bubbles to show your thinking. (Draw an area model on a separate sheet if it helps you.)

a. $1.38 \times 32 \approx$ _____ $\times$ _____ = _____ $1.38 \times 32 =$ _____

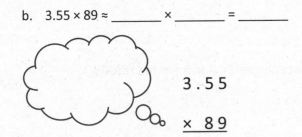

Think!
$1.38 \times 100 = 138$

$$1.38$$
$$\times \; 32$$

Think! 4,416 is 100 times too large! What is the real product?

$4,416 \div 100 = 44.16$

b. $3.55 \times 89 \approx$ _____ $\times$ _____ = _____ $3.55 \times 89 =$ _____

$$3.55$$
$$\times \; 89$$

EUREKA
MATH™

2. Solve using the standard algorithm.

   a. 5.04×8

   b. 147.83×67

   c. 83.41×504

   d. 0.56×432

3. Use the whole number product and place value reasoning to place the decimal point in the second product. Explain how you know.

   a. If $98 \times 768 = 75{,}264$ then $98 \times 7.68 =$ _____

   b. If $73 \times 1{,}563 = 114{,}099$ then $73 \times 15.63 =$ _____

   c. If $46 \times 1{,}239 = 56{,}994$ then $46 \times 123.9 =$ _____

Lesson 11:   Multiply decimal fractions by multi-digit whole numbers through conversion to a whole number problem and reasoning about the placement of the decimal.

EUREKA
MATH™

4. Jenny buys 22 pens that cost $1.15 each and 15 markers that cost $2.05 each. How much did Jenny spend?

5. A living room measures 24 feet by 15 feet. An adjacent square dining room measures 13 feet on each side. If carpet costs $6.98 per square foot, what is the total cost of putting carpet in both rooms?

Lesson 11:   Multiply decimal fractions by multi-digit whole numbers through conversion to a whole number problem and reasoning about the placement of the decimal.

59

EUREKA
MATH™

Name _____   Date _____

1. Estimate the product. Solve using the standard algorithm. Use the thought bubbles to show your thinking. (Draw an area model on a separate sheet if it helps you.)

    a.   2.42 × 12 ≈ _____ × _____ = _____   2.42 × 12 = _____

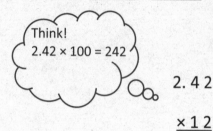

Think!
2.42 × 100 = 242

2. 4 2

× 1 2

Think! 2,904 is 100 times too large! What is the real product?

2,904 ÷ 100 = 29.04

    b.   4.13 × 37 ≈ _____ × _____ = _____   4.13 × 37 = _____

4. 1 3

× 3 7

Lesson 11:   Multiply decimal fractions by multi-digit whole numbers through conversion to a whole number problem and reasoning about the placement of the decimal.

EUREKA
MATH™

2. Solve using the standard algorithm.

    a. 2.03×13

    b. 53.16×34

    c. 371.23×53

    d. 1.57×432

3. Use the whole number product and place value reasoning to place the decimal point in the second product. Explain how you know.

    a.   If $36 \times 134 = 4{,}824$ then $36 \times 1.34 =$ _____

    b.   If $84 \times 2{,}674 = 224{,}616$ then $84 \times 26.74 =$ _____

    c. $19 \times 3{,}211 = 61{,}009$ then $321.1 \times 19 =$ _____

Lesson 11:   Multiply decimal fractions by multi-digit whole numbers through conversion to a whole number problem and reasoning about the placement of the decimal.

EUREKA MATH™

61

4. A slice of pizza costs $1.57. How much will 27 slices cost?

5. A spool of ribbon holds 6.75 meters. A craft club buys 21 spools.

   a. What is the total cost if the ribbon sells for $2 per meter?

   b. If the club uses 76.54 meters to complete a project, how much ribbon will be left?

Lesson 11:   Multiply decimal fractions by multi-digit whole numbers through conversion to a whole number problem and reasoning about the placement of the decimal.

EUREKA MATH™

Name _____ Date _____

1. Estimate. Then, solve using the standard algorithm. You may draw an area model if it helps you.

    a. $1.21 \times 14 \approx$ _____ $\times$ _____ = _____

    $$\begin{array}{r} 1.21 \\ \times\ \ 14 \\ \hline \end{array}$$

    b. $2.45 \times 305 \approx$ _____ $\times$ _____ = _____

    $$\begin{array}{r} 2.45 \\ \times\ 305 \\ \hline \end{array}$$

Lesson 12:   Reason about the product of a whole number and a decimal with hundredths using place value understanding and estimation.

63

©2015 Great Minds. eureka-math.org
G5-M2-SE-B1-1.3.1-01.2016

2. Estimate. Then, solve using the standard algorithm. Use a separate sheet to draw the area model if it helps you.

   a. $1.23 \times 12 \approx$ _____ × _____ = _____

   b. $1.3 \times 26 \approx$ _____ × _____ = _____

   c. $0.23 \times 14 \approx$ _____ × _____ = _____

   d. $0.45 \times 26 \approx$ _____ × _____ = _____

   e. $7.06 \times 28 \approx$ _____ × _____ = _____

   f. $6.32 \times 223 \approx$ _____ × _____ = _____

   g. $7.06 \times 208 \approx$ _____ × _____ = _____

   h. $151.46 \times 555 \approx$ _____ × _____ = _____

Lesson 12: Reason about the product of a whole number and a decimal with hundredths using place value understanding and estimation.

EUREKA MATH

3. Denise walks on the beach every afternoon. In the month of July, she walked 3.45 miles each day. How far did Denise walk during the month of July?

4. A gallon of gas costs $4.34. Greg puts 12 gallons of gas in his car. He has a 50-dollar bill. Tell how much money Greg will have left, or how much more money he will need. Show all your calculations.

5. Seth drinks a glass of orange juice every day that contains 0.6 grams of Vitamin C. He eats a serving of strawberries for snack after school every day that contains 0.35 grams of Vitamin C. How many grams of Vitamin C does Seth consume in 3 weeks?

EUREKA
MATH™

Lesson 12: Reason about the product of a whole number and a decimal with hundredths using place value understanding and estimation.

65

©2015 Great Minds. eureka-math.org
G5-M2-SE-B1-1.3.1-01.2016

Name _____   Date _____

1. Estimate. Then, solve using the standard algorithm. You may draw an area model if it helps you.

    a.   24 × 2.31 ≈ _____ × _____ = _____

$$\begin{array}{r} 2.31 \\ \times\ \ 24 \\ \hline \end{array}$$

    b.   5.42 × 305 ≈ _____ × _____ = _____

$$\begin{array}{r} 5.42 \\ \times\ 305 \\ \hline \end{array}$$

Lesson 12:   Reason about the product of a whole number and a decimal with hundredths using place value understanding and estimation.

©2015 Great Minds. eureka-math.org
G5-M2-SE-B1-1.3.1-01.2016

EUREKA MATH™

2. Estimate. Then, solve using the standard algorithm. Use a separate sheet to draw the area model if it helps you.

a. $1.23 \times 21 \approx$ _____ × _____ = _____

b. $3.2 \times 41 \approx$ _____ × _____ = _____

c. $0.32 \times 41 \approx$ _____ × _____ = _____

d. $0.54 \times 62 \approx$ _____ × _____ = _____

e. $6.09 \times 28 \approx$ _____ × _____ = _____

f. $6.83 \times 683 \approx$ _____ × _____ = _____

g. $6.09 \times 208 \approx$ _____ × _____ = _____

h. $171.76 \times 555 \approx$ _____ × _____ = _____

Lesson 12: Reason about the product of a whole number and a decimal with hundredths using place value understanding and estimation.

3. Eric's goal is to walk 2.75 miles to and from the park every day for an entire year. If he meets his goal, how many miles will Eric walk?

4. Art galleries often price paintings by the square inch. If a painting measures 22.5 inches by 34 inches and costs $4.15 per square inch, what is the selling price for the painting?

5. Gerry spends $1.25 each day on lunch at school. On Fridays, she buys an extra snack for $0.55. How much money will she spend in two weeks?

Lesson 12:   Reason about the product of a whole number and a decimal with hundredths using place value understanding and estimation.

Name _____ Date _____

1. Solve. The first one is done for you.

<table>
<tr>
<td>

a. Convert weeks to days.

8 weeks = 8 × (1 week)

= 8 × (7 days)

= 56 days

</td>
<td>

b. Convert years to days.

4 years = _____ × (_____ year)

= _____ × (_____ days)

= _____ days

</td>
</tr>
<tr>
<td>

c. Convert meters to centimeters.

9.2 m = _____ × (_____ m)

= _____ × (_____ cm)

= _____ cm

</td>
<td>

d. Convert yards to feet.

5.7 yards

</td>
</tr>
<tr>
<td>

e. Convert kilograms to grams.

6.08 kg

</td>
<td>

f. Convert pounds to ounces.

12.5 pounds

</td>
</tr>
</table>

EUREKA
MATH

Lesson 13: Use whole number multiplication to express equivalent measurements.

69

2. After solving, write a statement to express each conversion. The first one is done for you.

a. Convert the number of hours in a day to minutes. 24 hours = 24 × (1 hour) = 24 × (60 minutes) = 1,440 minutes One day has 24 hours, which is the same as 1,440 minutes.	b. A small female gorilla weighs 68 kilograms. How much does she weigh in grams?
c. The height of a man is 1.7 meters. What is his height in centimeters?	d. The capacity of a syringe is 0.08 liters. Convert this to milliliters.
e. A coyote weighs 11.3 pounds. Convert the coyote's weight to ounces.	f. An alligator is 2.3 yards long. What is the length of the alligator in inches?

Lesson 13:   Use whole number multiplication to express equivalent measurements.

Name _____   Date _____

1. Solve. The first one is done for you.

a. Convert weeks to days. 6 weeks = 6 × (1 week) = 6 × (7 days) = 42 days	b. Convert years to days. 7 years = _____ × (_____ year) = _____ × (_____ days) = _____ days
c. Convert meters to centimeters. 4.5 m = _____ × (_____ m) = _____ × (_____ cm) = _____ cm	d. Convert pounds to ounces. 12.6 pounds
e. Convert kilograms to grams. 3.09 kg	f. Convert yards to inches. 245 yd

 EUREKA MATH

Lesson 13:   Use whole number multiplication to express equivalent measurements.

71

2. After solving, write a statement to express each conversion. The first one is done for you.

a. Convert the number of hours in a day to minutes.
 	
   24 hours = 24 × (1 hour)
 	
                 = 24 × (60 minutes)
 	
                 = 1,440 minutes
 	
One day has 24 hours, which is the same as 1,440 minutes.
	b. A newborn giraffe weighs about 65 kilograms. How much does it weigh in grams?

c. The average height of a female giraffe is 4.6 meters. What is her height in centimeters?	d. The capacity of a beaker is 0.1 liter. Convert this to milliliters.
e. A pig weighs 9.8 pounds. Convert the pig's weight to ounces.	f. A marker is 0.13 meters long. What is the length in millimeters?

Lesson 13:   Use whole number multiplication to express equivalent measurements.

©2015 Great Minds. eureka-math.org
G5-M2-SE-B1-1.3.1-01.2016

0 cm
10 cm
20 cm

30 cm
40 cm
50 cm

60 cm
70 cm
80 cm

90 cm
100 cm

meter strip

Lesson 13: Use whole number multiplication to express equivalent measurements.

LEGEND   ------ CUT   ----- ALIGN EDGE

EUREKA
MATH™

©2015 Great Minds. eureka-math.org
G5-M2-SE-B1-1.3.1-01.2016

This page intentionally left blank

Name _____   Date _____

1. Solve. The first one is done for you.

a. Convert days to weeks.
 	
28 days = 28 × (1 day)
 	
     = 28 × ($\frac{1}{7}$ week)
 	
     = $\frac{28}{7}$ week
 	
     = 4 weeks
	b. Convert quarts to gallons.
	
	20 quarts = _____ × (1 quart)
	
	     = _____ × ($\frac{1}{4}$ gallon)
	
	     = _____ gallons
	
	     = _____ gallons

c. Convert centimeters to meters.
 	
920 cm = _____ × (_____ cm)
 	
     = _____ × (_____ m)
 	
     = _____ m
	d. Convert meters to kilometers.
	
	1,578 m = _____ × (_____ m)
	
	     = _____ × (0.001 km)
	
	     = _____ km

e. Convert grams to kilograms. 6,080 g =	**f. Convert milliliters to liters.** 509 mL =

2. After solving, write a statement to express each conversion. The first one is done for you.

a. The screen measures 24 inches. Convert 24 inches to feet. 24 inches = 24 × (1 inch) $= 24 \times \left(\frac{1}{12} \text{ feet}\right)$ $= \frac{24}{12}$ feet = 2 feet The screen measures 24 inches or 2 feet.	b. A jug of syrup holds 12 cups. Convert 12 cups to pints.
c. The length of the diving board is 378 centimeters. What is its length in meters?	d. The capacity of a container is 1,478 milliliters. Convert this to liters.
e. A truck weighs 3,900,000 grams. Convert the truck's weight to kilograms.	f. The distance was 264,040 meters. Convert the distance to kilometers.

Lesson 14:   Use fraction and decimal multiplication to express equivalent measurements.

EUREKA MATH™

Name _____ Date _____

1. Solve. The first one is done for you.

a. Convert days to weeks.	b. Convert quarts to gallons.
42 days = 42 × (1 day)	36 quarts = _____ × (1 quart)
$= 42 × \left(\frac{1}{7} \text{ week}\right)$	$= \underline{\hspace{1cm}} × \left(\frac{1}{4} \text{ gallon}\right)$
$= \frac{42}{7}$ week	= _____ gallons
= 6 weeks	= _____ gallons
c. Convert centimeters to meters.	d. Convert meters to kilometers.
760 cm = _____ × (_____ cm)	2,485 m = _____ × (_____ m)
= _____ × (_____ m)	= _____ × (0.001 km)
= _____ m	= _____ km
e. Convert grams to kilograms.	f. Convert milliliters to liters.
3,090 g =	205 mL =

EUREKA
MATH™

Lesson 14: Use fraction and decimal multiplication to express equivalent measurements.

77

©2015 Great Minds. eureka-math.org
G5-M2-SE-B1-1.3.1-01.2016

2. After solving, write a statement to express each conversion. The first one is done for you.

a. The screen measures 36 inches. Convert 36 inches to feet. $36 \text{ inches} = 36 \times (1 \text{ inch})$ $= 36 \times \left(\frac{1}{12} \text{ feet}\right)$ $= \frac{36}{12} \text{ feet}$ $= 3 \text{ feet}$ The screen measures 36 inches or 3 feet.	b. A jug of juice holds 8 cups. Convert 8 cups to pints.
c. The length of the flower garden is 529 centimeters. What is its length in meters?	d. The capacity of a container is 2,060 milliliters. Convert this to liters.
e. A hippopotamus weighs 1,560,000 grams. Convert the hippopotamus' weight to kilograms.	f. The distance was 372,060 meters. Convert the distance to kilometers.

Lesson 14: Use fraction and decimal multiplication to express equivalent measurements.

EUREKA MATH™

Name _____   Date _____

Solve.

1. Liza's cat had six kittens! When Liza and her brother weighed all the kittens together, they weighed 4 pounds 2 ounces. Since all the kittens are about the same size, about how many ounces does each kitten weigh?

2. A container of oregano is 17 pounds heavier than a container of peppercorns. Their total weight is 253 pounds. The peppercorns will be sold in one-ounce bags. How many bags of peppercorns can be made?

Lesson 15:   Solve two-step word problems involving measurement conversions.

79

3. Each costume needs 46 centimeters of red ribbon and 3 times as much yellow ribbon. What is the total length of ribbon needed for 64 costumes? Express your answer in meters.

4. When making a batch of orange juice for her basketball team, Jackie used 5 times as much water as concentrate. There were 32 more cups of water than concentrate.

   a. How much juice did she make in all?

   b. She poured the juice into quart containers. How many containers could she fill?

Lesson 15:   Solve two-step word problems involving measurement conversions.

Name _____   Date _____

Solve.

1. Tia cut a 4-meter 8-centimeter wire into 10 equal pieces.   Marta cut a 540-centimeter wire into 9 equal pieces. How much longer is one of Marta's wires than one of Tia's?

2. Jay needs 19 quarts more paint for the outside of his barn than for the inside. If he uses 107 quarts in all, how many gallons of paint will be used to paint the inside of the barn?

3. String A is 35 centimeters long. String B is 5 times as long as String A. Both are necessary to create a decorative bottle. Find the total length of string needed for 17 identical decorative bottles. Express your answer in meters.

4. A pineapple is 7 times as heavy as an orange. The pineapple also weighs 870 grams more than the orange.

    a. What is the total weight in grams for the pineapple and orange?

    b. Express the total weight of the pineapple and orange in kilograms.

Lesson 15:   Solve two-step word problems involving measurement conversions.

EUREKA
MATH

Name _____   Date _____

1. Divide. Draw place value disks to show your thinking for (a) and (c). You may draw disks on your personal white board to solve the others if necessary.

a. 500 ÷ 10	b. 360 ÷ 10
c. 12,000 ÷ 100	d. 450,000 ÷ 100
e. 700,000 ÷ 1,000	f. 530,000 ÷ 100

2. Divide. The first one is done for you.

a. 12,000 ÷ 30 = 12,000 ÷ 10 ÷ 3 = 1,200 ÷ 3 = 400	b. 12,000 ÷ 300	c. 12,000 ÷ 3,000
d. 560,000 ÷ 70	e. 560,000 ÷ 700	f. 560,000 ÷ 7,000
g. 28,000 ÷ 40	h. 450,000 ÷ 500	i. 810,000 ÷ 9,000

Lesson 16:   Use *divide by 10* patterns for multi-digit whole number division.

3. The floor of a rectangular banquet hall has an area of 3,600 m². The length is 90 m.

   a. What is the width of the banquet hall?

   b. A square banquet hall has the same area. What is the length of the room?

   c. A third rectangular banquet hall has a perimeter of 3,600 m. What is the width if the length is 5 times the width?

©2015 Great Minds. eureka-math.org
G5-M2-SE-B1-1.3.1-01.2016

4. Two fifth graders solved 400,000 divided by 800. Carter said the answer is 500, while Kim said the answer is 5,000.

   a. Who has the correct answer? Explain your thinking.

   b. What if the problem is 4,000,000 divided by 8,000? What is the quotient?

Lesson 16:   Use *divide by 10* patterns for multi-digit whole number division.

EUREKA
MATH™

Name _____ Date _____

1. Divide. Draw place value disks to show your thinking for (a) and (c). You may draw disks on your personal white board to solve the others if necessary.

a. 300 ÷ 10	b. 450 ÷ 10
c. 18,000 ÷ 100	d. 730,000 ÷ 100
e. 900,000 ÷ 1,000	f. 680,000 ÷ 1,000

2. Divide. The first one is done for you.

a.   18,000 ÷ 20
 	
    = 18,000 ÷ 10 ÷ 2
 	
    = 1,800 ÷ 2
 	
    = 900
	b. 18,000 ÷ 200	c. 18,000 ÷ 2,000

d. 420,000 ÷ 60	e. 420,000 ÷ 600	f. 420,000 ÷ 6,000
g. 24,000 ÷ 30	h. 560,000 ÷ 700	i. 450,000 ÷ 9,000

Lesson 16:   Use *divide by 10* patterns for multi-digit whole number division.

©2015 Great Minds. eureka-math.org
 G5-M2-SE-B1-1.3.1-01.2016

3. A stadium holds 50,000 people. The stadium is divided into 250 different seating sections. How many seats are in each section?

4. Over the course of a year, a tractor trailer commutes 160,000 miles across America.

   a. Assuming a trucker changes his tires every 40,000 miles, and that he starts with a brand new set of tires, how many sets of tires will he use in a year?

   b. If the trucker changes the oil every 10,000 miles, and he starts the year with a fresh oil change, how many times will he change the oil in a year?

EUREKA
MATH™

Lesson 16: Use *divide by 10* patterns for multi-digit whole number division.

89

©2015 Great Minds. eureka-math.org
G5-M2-SE-B1-1.3.1-01.2016

This page intentionally left blank

Name _____ Date _____

1. Estimate the quotient for the following problems. Round the divisor first.

a. $609 \div 21$ $\approx 600 \div 20$ $= 30$	b. $913 \div 29$ $\approx$ _____ $\div$ _____ $=$ _____	c. $826 \div 37$ $\approx$ _____ $\div$ _____ $=$ _____
d. $141 \div 73$ $\approx$ _____ $\div$ _____ $=$ _____	e. $241 \div 58$ $\approx$ _____ $\div$ _____ $=$ _____	f. $482 \div 62$ $\approx$ _____ $\div$ _____ $=$ _____
g. $656 \div 81$ $\approx$ _____ $\div$ _____ $=$ _____	h. $799 \div 99$ $\approx$ _____ $\div$ _____ $=$ _____	i. $635 \div 95$ $\approx$ _____ $\div$ _____ $=$ _____
j. $311 \div 76$ $\approx$ _____ $\div$ _____ $=$ _____	k. $648 \div 83$ $\approx$ _____ $\div$ _____ $=$ _____	l. $143 \div 35$ $\approx$ _____ $\div$ _____ $=$ _____
m. $525 \div 25$ $\approx$ _____ $\div$ _____ $=$ _____	n. $552 \div 85$ $\approx$ _____ $\div$ _____ $=$ _____	o. $667 \div 11$ $\approx$ _____ $\div$ _____ $=$ _____

Lesson 17: Use basic facts to approximate quotients with two-digit divisors.

91

EUREKA MATH™

2. A video game store has a budget of $825, and would like to purchase new video games. If each video game costs $41, estimate the total number of video games the store can purchase with its budget. Explain your thinking.

3. Jackson estimated 637 ÷ 78 as 640 ÷ 80. He reasoned that 64 tens divided by 8 tens should be 8 tens. Is Jackson's reasoning correct? If so, explain why. If not, explain a correct solution.

Lesson 17:   Use basic facts to approximate quotients with two-digit divisors.

Name _____   Date _____

1. Estimate the quotient for the following problems. The first one is done for you.

a. $821 \div 41$ $\approx 800 \div 40$ $= 20$	b. $617 \div 23$ $\approx$ _____ $\div$ _____ $=$ _____	c. $821 \div 39$ $\approx$ _____ $\div$ _____ $=$ _____
d. $482 \div 52$ $\approx$ _____ $\div$ _____ $=$ _____	e. $531 \div 48$ $\approx$ _____ $\div$ _____ $=$ _____	f. $141 \div 73$ $\approx$ _____ $\div$ _____ $=$ _____
g. $476 \div 81$ $\approx$ _____ $\div$ _____ $=$ _____	h. $645 \div 69$ $\approx$ _____ $\div$ _____ $=$ _____	i. $599 \div 99$ $\approx$ _____ $\div$ _____ $=$ _____
j. $301 \div 26$ $\approx$ _____ $\div$ _____ $=$ _____	k. $729 \div 81$ $\approx$ _____ $\div$ _____ $=$ _____	l. $636 \div 25$ $\approx$ _____ $\div$ _____ $=$ _____
m. $835 \div 89$ $\approx$ _____ $\div$ _____ $=$ _____	n. $345 \div 72$ $\approx$ _____ $\div$ _____ $=$ _____	o. $559 \div 11$ $\approx$ _____ $\div$ _____ $=$ _____

2. Mrs. Johnson spent $611 buying lunch for 78 students. If all the lunches cost the same, about how much did she spend on each lunch?

3. An oil well produces 172 gallons of oil every day. A standard oil barrel holds 42 gallons of oil. About how many barrels of oil will the well produce in one day? Explain your thinking.

Lesson 17:   Use basic facts to approximate quotients with two-digit divisors.

©2015 Great Minds. eureka-math.org
G5-M2-SE-B1-1.3.1-01.2016

Name _____   Date _____

1. Estimate the quotients for the following problems. The first one is done for you.

a. 5,738 ÷ 21 ≈ 6,000 ÷ 20 = 300	b. 2,659 ÷ 28 ≈ _____ ÷ _____ = _____	c. 9,155 ÷ 34 ≈ _____ ÷ _____ = _____
d. 1,463 ÷ 53 ≈ _____ ÷ _____ = _____	e. 2,525 ÷ 64 ≈ _____ ÷ _____ = _____	f. 2,271 ÷ 72 ≈ _____ ÷ _____ = _____
g. 4,901 ÷ 75 ≈ _____ ÷ _____ = _____	h. 8,515 ÷ 81 ≈ _____ ÷ _____ = _____	i. 8,515 ÷ 89 ≈ _____ ÷ _____ = _____
j. 3,925 ÷ 68 ≈ _____ ÷ _____ = _____	k. 5,124 ÷ 81 ≈ _____ ÷ _____ = _____	l. 4,945 ÷ 93 ≈ _____ ÷ _____ = _____
m. 5,397 ÷ 94 ≈ _____ ÷ _____ = _____	n. 6,918 ÷ 86 ≈ _____ ÷ _____ = _____	o. 2,806 ÷ 15 ≈ _____ ÷ _____ = _____

Lesson 18: Use basic facts to approximate quotients with two-digit divisors.

95

©2015 Great Minds. eureka-math.org
G5-M2-SE-B1-1.3.1-01.2016

2. A swimming pool requires 672 ft² of floor space. The length of the swimming pool is 32 ft. Estimate the width of the swimming pool.

3. Janice bought 28 apps for her phone that, altogether, used 1,348 MB of space.

    a. If each app used the same amount of space, about how many MB of memory did each app use?
        Show how you estimated.

    b. If half of the apps were free and the other half were $1.99 each, about how much did she spend?

4. A quart of paint covers about 85 square feet. About how many quarts would you need to cover a fence with an area of 3,817 square feet?

5. Peggy has saved $9,215. If she is paid $45 an hour, about how many hours did she work?

Lesson 18:   Use basic facts to approximate quotients with two-digit divisors.

EUREKA
MATH™

Name _____   Date _____

1. Estimate the quotients for the following problems. The first one is done for you.

a. 8,328 ÷ 41 ≈ 8,000 ÷ 40 = 200	b. 2,109 ÷ 23 ≈ _____ ÷ _____ = _____	c. 8,215 ÷ 38 ≈ _____ ÷ _____ = _____
d. 3,861 ÷ 59 ≈ _____ ÷ _____ = _____	e. 2,899 ÷ 66 ≈ _____ ÷ _____ = _____	f. 5,576 ÷ 92 ≈ _____ ÷ _____ = _____
g. 5,086 ÷ 73 ≈ _____ ÷ _____ = _____	h. 8,432 ÷ 81 ≈ _____ ÷ _____ = _____	i. 9,032 ÷ 89 ≈ _____ ÷ _____ = _____
j. 2,759 ÷ 48 ≈ _____ ÷ _____ = _____	k. 8,194 ÷ 91 ≈ _____ ÷ _____ = _____	l. 4,368 ÷ 63 ≈ _____ ÷ _____ = _____
m. 6,537 ÷ 74 ≈ _____ ÷ _____ = _____	n. 4,998 ÷ 48 ≈ _____ ÷ _____ = _____	o. 6,106 ÷ 25 ≈ _____ ÷ _____ = _____

EUREKA MATH™

Lesson 18:   Use basic facts to approximate quotients with two-digit divisors.

97

2. 91 boxes of apples hold a total of 2,605 apples. Assuming each box has about the same number of apples, estimate the number of apples in each box.

3. A wild tiger can eat up to 55 pounds of meat in a day. About how many days would it take for a tiger to eat the following prey?

Prey	Weight of Prey	Number of Days
Eland Antelope	1,754 pounds	
Boar	661 pounds	
Chital Deer	183 pounds	
Water Buffalo	2,322 pounds	

EUREKA
MATH™

Name _____   Date _____

1. Divide, and then check. The first problem is done for you.

a. 41 ÷ 30

$$
\begin{array}{r}
1 \ \text{R } 11 \\
3\;0\,\overline{\big)\,4\;\;1} \\
-\underline{\;3\;\;0} \\
1\;\;1
\end{array}
$$

Check:

30 × 1 = 30
30 + 11 = 41

b. 80 ÷ 30

c. 71 ÷ 50

d. 270 ÷ 30

e. 643 ÷ 80

f. 215 ÷ 90

EUREKA
MATH™

2. Terry says the solution to 299 ÷ 40 is 6 with a remainder of 59. His work is shown below. Explain Terry's error in thinking, and then find the correct quotient using the space on the right.

```
            6
   4 0 | 2 9 9
     -  2 4 0
            5 9
```

```
   4 0 | 2 9 9
```

3. A number divided by 80 has a quotient of 7 with 4 as a remainder. Find the number.

4. While swimming a 2 km race, Adam changes from breaststroke to butterfly every 200 m. How many times does he switch strokes during the first half of the race?

Lesson 19: Divide two- and three-digit dividends by multiples of 10 with single-digit quotients, and make connections to a written method.

EUREKA MATH

©2015 Great Minds. eureka-math.org
G5-M2-SE-B1-1.3.1-01.2016

Name _____   Date _____

1. Divide, and then check using multiplication. The first one is done for you.

    a. 71 ÷ 20

$$
\begin{array}{r}
3 \quad \text{R } 11 \\
2\ 0\ \overline{)\ 7\quad 1} \\
-\ 6\quad 0 \\
\hline
1\quad 1
\end{array}
$$

    Check:

    20 × 3 = 60
    60 + 11 = 71

    b. 90 ÷ 40

    c. 95 ÷ 60

    d. 280 ÷ 30

    e. 437 ÷ 60

    f. 346 ÷ 80

EUREKA
MATH™

Lesson 19:   Divide two- and three-digit dividends by multiples of 10 with
single-digit quotients, and make connections to a written method.

101

©2015 Great Minds. eureka-math.org
G5-M2-SE-B1-1.3.1-01.2016

2. A number divided by 40 has a quotient of 6 with a remainder of 16. Find the number.

3. A shipment of 288 reams of paper was delivered. Each of the 30 classrooms received an equal share of the paper. Any extra reams of paper were stored. After the paper was distributed to the classrooms, how many reams of paper were stored?

4. How many groups of sixty are in two hundred forty-four?

Lesson 19:   Divide two- and three-digit dividends by multiples of 10 with single-digit quotients, and make connections to a written method.

©2015 Great Minds. eureka-math.org
G5-M2-SE-B1-1.3.1-01.2016

Name _____   Date _____

1. Divide. Then, check with multiplication. The first one is done for you.

   a. $65 \div 17$ b. $49 \div 21$

```
            3 R 14      Check:
    17   6 5
     -   5 1            17 × 3 = 51
         1 4            51 + 14 = 65
```

   c. $78 \div 39$ d. $84 \div 32$

   e. $77 \div 25$ f. $68 \div 17$

Lesson 20:   Divide two- and three-digit dividends by two-digit divisors with single-digit quotients, and make connections to a written method.

©2015 Great Minds. eureka-math.org
G5-M2-SE-B1-1.3.1-01.2016

2. When dividing 82 by 43, Linda estimated the quotient to be 2. Examine Linda's work, and explain what she needs to do next. On the right, show how you would solve the problem.

Linda's Estimation:	Linda's Work:	Your Work:

```
        2                    2
40 | 8 0            43 | 8 2          43 | 8 2
                       - 8 6
                         ? ?
```

3. A number divided by 43 has a quotient of 3 with 28 as a remainder. Find the number. Show your work.

Lesson 20: Divide two- and three-digit dividends by two-digit divisors with single-digit quotients, and make connections to a written method.

EUREKA MATH

4. Write another division problem that has a quotient of 3 and a remainder of 28.

5. Mrs. Silverstein sold 91 cupcakes at a food fair. The cupcakes were sold in boxes of "a baker's dozen," which is 13. She sold all the cupcakes at $15 per box. How much money did she receive?

Lesson 20: Divide two- and three-digit dividends by two-digit divisors with single-digit quotients, and make connections to a written method.

105

©2015 Great Minds. eureka-math.org
G5-M2-SE-B1-1.3.1-01.2016

Name _____   Date _____

1. Divide. Then, check with multiplication. The first one is done for you.

   a. $72 \div 31$

   $$\begin{array}{r} 2 \ \text{R} \ 10 \\ 31\overline{\smash{)}72} \\ -\ 62 \\ \hline 10 \end{array}$$

   Check:

   $31 \times 2 = 62$

   $62 + 10 = 72$

   b. $89 \div 21$

   c. $94 \div 33$

   d. $67 \div 19$

   e. $79 \div 25$

   f. $83 \div 21$

Lesson 20:   Divide two- and three-digit dividends by two-digit divisors with single-digit quotients, and make connections to a written method.

EUREKA MATH

©2015 Great Minds. eureka-math.org
G5-M2-SE-B1-1.3.1-01.2016

2. A 91 square foot bathroom has a length of 13 feet. What is the width of the bathroom?

3. While preparing for a morning conference, Principal Corsetti is laying out 8 dozen bagels on square plates. Each plate can hold 14 bagels.

    a. How many plates of bagels will Mr. Corsetti have?

    b. How many more bagels would be needed to fill the final plate with bagels?

This page intentionally left blank

Name _____   Date _____

1. Divide. Then, check using multiplication. The first one is done for you.

    a. 258 ÷ 47

                          5 R 23
                  4 7 | 2 5 8
                    -   2 3 5
                    _____
                        2 3

    Check:

    47 × 5 = 235

    235 + 23 = 258

    b. 148 ÷ 67

    c. 591 ÷ 73

    d. 759 ÷ 94

e. $653 \div 74$

f. $257 \div 36$

2. Generate and solve at least one more division problem with the same quotient and remainder as the one below. Explain your thought process.

$$
\begin{array}{r}
8 \\
58\ \overline{\smash{\big)}\ 4\ 7\ 5} \\
-\ 4\ 6\ 4 \\
\hline
1\ 1
\end{array}
$$

Lesson 21:   Divide two- and three-digit dividends by two-digit divisors with single-digit quotients, and make connections to a written method.

EUREKA
MATH™

3. Assume that Mrs. Giang's car travels 14 miles on each gallon of gas. If she travels to visit her niece who lives 133 miles away, how many gallons of gas will Mrs. Giang need to make the round trip?

4. Louis brings 79 pencils to school. After he gives each of his 15 classmates an equal number of pencils, he will give any leftover pencils to his teacher.

   a. How many pencils will Louis's teacher receive?

   b. If Louis decides instead to take an equal share of the pencils along with his classmates, will his teacher receive more pencils or fewer pencils? Show your thinking.

Lesson 21:   Divide two- and three-digit dividends by two-digit divisors with single-digit quotients, and make connections to a written method.

111

Name _____   Date _____

1. Divide. Then, check using multiplication. The first one is done for you.

   a. 129 ÷ 21

   $$\begin{array}{r} 6\ R\ 3 \\ 21\overline{\smash{)}1\ 2\ 9} \\ -\ \underline{1\ 2\ 6} \\ 3 \end{array}$$

   Check:

   21 × 6 = 126

   126 + 3 = 129

   b. 158 ÷ 37

   c. 261 ÷ 49

   d. 574 ÷ 82

Lesson 21: Divide two- and three-digit dividends by two-digit divisors with single-digit quotients, and make connections to a written method.

EUREKA MATH

e. $464 \div 58$

f. $640 \div 79$

2. It takes Juwan exactly 35 minutes by car to get to his grandmother's. The nearest parking area is a 4-minute walk from her apartment. One week, he realized that he spent 5 hours and 12 minutes traveling to her apartment and then back home. How many round trips did he make to visit his grandmother?

EUREKA
MATH™

Lesson 21:   Divide two- and three-digit dividends by two-digit divisors with single-digit quotients, and make connections to a written method.

113

3.   How many eighty-fours are in 672?

Lesson 21:   Divide two- and three-digit dividends by two-digit divisors with single-
digit quotients, and make connections to a written method.

EUREKA
MATH™

Name _____ Date _____

1. Divide. Then, check using multiplication. The first one is done for you.

a. 580 ÷ 17

```
          3  4 R 2
   17 | 5  8  0
      -  5  1
          7  0
       -  6  8
             2
```

Check:

34 × 17 = 578

578 + 2 = 580

b. 730 ÷ 32

c. 940 ÷ 28

d. 553 ÷ 23

Lesson 22: Divide three- and four-digit dividends by two-digit divisors resulting in two- and three-digit quotients, reasoning about the decomposition of successive remainders in each place value.

115

EUREKA MATH™

©2015 Great Minds. eureka-math.org
G5-M2-SE-B1-1.3.1-01.2016

e. $704 \div 46$

f. $614 \div 15$

2. Halle solved $664 \div 48$ below. She got a quotient of 13 with a remainder of 40. How could she use her work below to solve $659 \div 48$ without redoing the work? Explain your thinking.

```
           1 3
    4 8 | 6 6 4
      -   4 8
          1 8 4
      - 1 4 4
            4 0
```

Lesson 22:   Divide three- and four-digit dividends by two-digit divisors resulting in two- and three-digit quotients, reasoning about the decomposition of successive remainders in each place value.

EUREKA
MATH™

3. 27 students are learning to make balloon animals. There are 172 balloons to be shared equally among the students.

    a. How many balloons are left over after sharing them equally?

    b. If each student needs 7 balloons, how many more balloons are needed? Explain how you know.

Lesson 22:   Divide three- and four-digit dividends by two-digit divisors resulting in   117
two- and three-digit quotients, reasoning about the decomposition of
successive remainders in each place value.

©2015 Great Minds. eureka-math.org
G5-M2-SE-B1-1.3.1-01.2016

Name _____ Date _____

1. Divide. Then, check using multiplication. The first one is done for you.

   a. $487 \div 21$

   $$
   \begin{array}{r}
   2\ 3\ \text{R}\ 4 \\
   21\ \overline{)\ 4\ 8\ 7} \\
   -\ 4\ 2 \\
   \hline
   6\ 7 \\
   -\ 6\ 3 \\
   \hline
   4
   \end{array}
   $$

   Check:

   $21 \times 23 = 483$

   $483 + 4 = 487$

   b. $485 \div 15$

   c. $700 \div 21$

   d. $399 \div 31$

Lesson 22: Divide three- and four-digit dividends by two-digit divisors resulting in two- and three-digit quotients, reasoning about the decomposition of successive remainders in each place value.

EUREKA MATH

e. 820 ÷ 42

f. 908 ÷ 56

2. When dividing 878 by 31, a student finds a quotient of 28 with a remainder of 11. Check the student's work, and use the check to find the error in the solution.

Lesson 22:   Divide three- and four-digit dividends by two-digit divisors resulting in two- and three-digit quotients, reasoning about the decomposition of successive remainders in each place value.

119

3. A baker was going to arrange 432 desserts into rows of 28. The baker divides 432 by 28 and gets a quotient of 15 with remainder 12. Explain what the quotient and remainder represent.

Lesson 22:   Divide three- and four-digit dividends by two-digit divisors resulting in two- and three-digit quotients, reasoning about the decomposition of successive remainders in each place value.

©2015 Great Minds. eureka-math.org
G5-M2-SE-B1-1.3.1-01.2016

Name _____   Date _____

1. Divide. Then, check using multiplication.

a.   4,859 ÷ 23

b.   4,368 ÷ 52

c.   7,242 ÷ 34

d.   3,164 ÷ 45

e.   9,152 ÷ 29

f.   4,424 ÷ 63

Lesson 23:   Divide three- and four-digit dividends by two-digit divisors resulting in
two- and three-digit quotients, reasoning about the decomposition of
successive remainders in each place value.

121

2. Mr. Riley baked 1,692 chocolate cookies. He sold them in boxes of 36 cookies each. How much money did he collect if he sold them all at $8 per box?

3. 1,092 flowers are arranged into 26 vases, with the same number of flowers in each vase. How many flowers would be needed to fill 130 such vases?

4. The elephant's water tank holds 2,560 gallons of water. After two weeks, the zookeeper measures and finds that the tank has 1,944 gallons of water left. If the elephant drinks the same amount of water each day, how many days will a full tank of water last?

Lesson 23:   Divide three- and four-digit dividends by two-digit divisors resulting in two- and three-digit quotients, reasoning about the decomposition of successive remainders in each place value.

Name _____ Date _____

1. Divide. Then, check using multiplication.

   a. $9,962 \div 41$

   b. $1,495 \div 45$

   c. $6,691 \div 28$

   d. $2,625 \div 32$

   e. $2,409 \div 19$

   f. $5,821 \div 62$

Lesson 23: Divide three- and four-digit dividends by two-digit divisors resulting in two- and three-digit quotients, reasoning about the decomposition of successive remainders in each place value.

123

EUREKA MATH

2. A political gathering in South America was attended by 7,910 people. Each of South America's 14 countries was equally represented. How many representatives attended from each country?

3. A candy company packages caramel into containers that hold 32 fluid ounces. In the last batch, 1,848 fluid ounces of caramel were made. How many containers were needed for this batch?

Lesson 23: Divide three- and four-digit dividends by two-digit divisors resulting in two- and three-digit quotients, reasoning about the decomposition of successive remainders in each place value.

EUREKA MATH™

Name _____ Date _____

1. Divide. Show the division in the right-hand column in two steps. The first two have been done for you.

   a. $1.2 \div 6 = 0.2$

   b. $1.2 \div 60 = (1.2 \div 6) \div 10 = 0.2 \div 10 = 0.02$

   c. $2.4 \div 4 =$ _____

   d. $2.4 \div 40 =$ _____

   e. $14.7 \div 7 =$ _____

   f. $14.7 \div 70 =$ _____

   g. $0.34 \div 2 =$ _____

   h. $3.4 \div 20 =$ _____

   i. $0.45 \div 9 =$ _____

   j. $0.45 \div 90 =$ _____

   k. $3.45 \div 3 =$ _____

   l. $34.5 \div 300 =$ _____

Lesson 24: Divide decimal dividends by multiples of 10, reasoning about the placement of the decimal point and making connections to a written method.

125

EUREKA MATH™

2. Use place value reasoning and the first quotient to compute the second quotient. Explain your thinking.

a. $46.5 \div 5 = 9.3$

$46.5 \div 50 = $ _____

b. $0.51 \div 3 = 0.17$

$0.51 \div 30 = $ _____

c. $29.4 \div 70 = 0.42$

$29.4 \div 7 = $ _____

d. $13.6 \div 40 = 0.34$

$13.6 \div 4 = $ _____

Lesson 24: Divide decimal dividends by multiples of 10, reasoning about the placement of the decimal point and making connections to a written method.

EUREKA
MATH™

3. Twenty polar bears live at the zoo. In four weeks, they eat 9,732.8 pounds of food altogether. Assuming each bear is fed the same amount of food, how much food is used to feed one bear for a week? Round your answer to the nearest pound.

4. The total weight of 30 bags of flour and 4 bags of sugar is 42.6 kg. If each bag of sugar weighs 0.75 kg, what is the weight of each bag of flour?

Lesson 24:   Divide decimal dividends by multiples of 10, reasoning about the
placement of the decimal point and making connections to a written
method.

127

©2015 Great Minds. eureka-math.org
G5-M2-SE-B1-1.3.1-01.2016

Name _____ Date _____

1. Divide. Show every other division sentence in two steps. The first two have been done for you.

   a. $1.8 \div 6 = 0.3$

   b. $1.8 \div 60 = (1.8 \div 6) \div 10 = 0.3 \div 10 = 0.03$

   c. $2.4 \div 8 = $ _____

   d. $2.4 \div 80 = $ _____

   e. $14.6 \div 2 = $ _____

   f. $14.6 \div 20 = $ _____

   g. $0.8 \div 4 = $ _____

   h. $80 \div 400 = $ _____

   i. $0.56 \div 7 = $ _____

   j. $0.56 \div 70 = $ _____

   k. $9.45 \div 9 = $ _____

   l. $9.45 \div 900 = $ _____

Lesson 24: Divide decimal dividends by multiples of 10, reasoning about the placement of the decimal point and making connections to a written method.

EUREKA MATH

2. Use place value reasoning and the first quotient to compute the second quotient. Use place value to explain how you placed the decimal point.

   a. $65.6 \div 80 = 0.82$

   $65.6 \div 8 =$ _____

   b. $2.5 \div 50 = 0.05$

   $2.5 \div 5 =$ _____

   c. $19.2 \div 40 = 0.48$

   $19.2 \div 4 =$ _____

   d. $39.6 \div 6 = 6.6$

   $39.6 \div 60 =$ _____

Lesson 24:   Divide decimal dividends by multiples of 10, reasoning about the
placement of the decimal point and making connections to a written
method.

129

©2015 Great Minds. eureka-math.org
G5-M2-SE-B1-1.3.1-01.2016

3. Chris rode his bike along the same route every day for 60 days. He logged that he had gone exactly 127.8 miles.

   a. How many miles did he bike each day? Show your work to explain how you know.

   b. How many miles did he bike over the course of two weeks?

4. 2.1 liters of coffee were equally distributed to 30 cups. How many milliliters of coffee were in each cup?

Lesson 24:   Divide decimal dividends by multiples of 10, reasoning about the placement of the decimal point and making connections to a written method.

Name _____   Date _____

1. Estimate the quotients.

    a. 3.24 ÷ 82 ≈

    b. 361.2 ÷ 61 ≈

    c. 7.15 ÷ 31 ≈

    d. 85.2 ÷ 31 ≈

    e. 27.97 ÷ 28 ≈

2. Estimate the quotient in (a). Use your estimated quotient to estimate (b) and (c).

    a. 7.16 ÷ 36 ≈

    b. 716 ÷ 36 ≈

    c. 71.6 ÷ 36 ≈

3. Edward bikes the same route to and from school each day. After 28 school days, he bikes a total distance of 389.2 miles.

   a. Estimate how many miles he bikes in one day.

   b. If Edward continues his routine of biking to school, about how many days altogether will it take him to reach a total distance of 500 miles?

4. Xavier goes to the store with $40. He spends $38.60 on 13 bags of popcorn.

   a. About how much does one bag of popcorn cost?

   b. Does he have enough money for another bag? Use your estimate to explain your answer.

Lesson 25:   Use basic facts to approximate decimal quotients with two-digit divisors, reasoning about the placement of the decimal point.

©2015 Great Minds. eureka-math.org
G5-M2-SE-B1-1.3.1-01.2016

EUREKA
MATH™

Name _____   Date _____

1. Estimate the quotients.

    a. $3.53 \div 51 \approx$

    b. $24.2 \div 42 \approx$

    c. $9.13 \div 23 \approx$

    d. $79.2 \div 39 \approx$

    e. $7.19 \div 58 \approx$

2. Estimate the quotient in (a). Use your estimated quotient to estimate (b) and (c).

    a. $9.13 \div 42 \approx$

    b. $913 \div 42 \approx$

    c. $91.3 \div 42 \approx$

3. Mrs. Huynh bought a bag of 3 dozen toy animals as party favors for her son's birthday party. The bag of toy animals cost $28.97. Estimate the price of each toy animal.

4. Carter drank 15.75 gallons of water in 4 weeks. He drank the same amount of water each day.

   a. Estimate how many gallons he drank in one day.

   b. Estimate how many gallons he drank in one week.

   c. About how many days altogether will it take him to drink 20 gallons?

Lesson 25: Use basic facts to approximate decimal quotients with two-digit divisors, reasoning about the placement of the decimal point.

Name _____   Date _____

1. 156 ÷ 24 and 102 ÷ 15 both have a quotient of 6 and a remainder of 12.

    a. Are the division expressions equivalent to each other? Use your knowledge of decimal division to justify your answer.

    b. Construct your own division problem with a two-digit divisor that has a quotient of 6 and a remainder of 12 but is not equivalent to the problems in 1(a).

2. Divide. Then, check your work with multiplication.

    a. 36.14 ÷ 13

    b. 62.79 ÷ 23

    c. 12.21 ÷ 11

    d. 6.89 ÷ 13

EUREKA
MATH™

Lesson 26:   Divide decimal dividends by two-digit divisors, estimating quotients, reasoning about the placement of the decimal point, and making connections to a written method.

135

e. $249.6 \div 52$

f. $24.96 \div 52$

g. $300.9 \div 59$

h. $30.09 \div 59$

3.   The weight of 72 identical marbles is 183.6 grams. What is the weight of each marble? Explain how you know the decimal point of your quotient is placed reasonably.

Lesson 26:   Divide decimal dividends by two-digit divisors, estimating quotients, reasoning about the placement of the decimal point, and making connections to a written method.

EUREKA MATH™

©2015 Great Minds. eureka-math.org
G5-M2-SE-B1-1.3.1-01.2016

4. Cameron wants to measure the length of his classroom using his foot as a length unit. His teacher tells him the length of the classroom is 23 meters. Cameron steps across the classroom heel to toe and finds that it takes him 92 steps. How long is Cameron's foot in meters?

5. A blue rope is three times as long as a red rope. A green rope is 5 times as long as the blue rope. If the total length of the three ropes is 508.25 meters, what is the length of the blue rope?

Lesson 26: Divide decimal dividends by two-digit divisors, estimating quotients, reasoning about the placement of the decimal point, and making connections to a written method.

137

©2015 Great Minds. eureka-math.org
G5-M2-SE-B1-1.3.1-01.2016

Name _____   Date _____

1. Create two whole number division problems that have a quotient of 9 and a remainder of 5. Justify which is greater using decimal division.

2. Divide. Then, check your work with multiplication.

    a. $75.9 \div 22$ b. $97.28 \div 19$

    c. $77.14 \div 38$ d. $12.18 \div 29$

Lesson 26:   Divide decimal dividends by two-digit divisors, estimating quotients, reasoning about the placement of the decimal point, and making connections to a written method.

EUREKA MATH

3. Divide.

   a. 97.58 ÷ 34

   b. 55.35 ÷ 45

4. Use the equations on the left to solve the problems on the right. Explain how you decided where to place the decimal in the quotient.

   a. 520.3 ÷ 43 = 12.1

   52.03 ÷ 43 = _____

   b. 19.08 ÷ 36 = 0.53

   190.8 ÷ 36 = _____

Lesson 26: Divide decimal dividends by two-digit divisors, estimating quotients, reasoning about the placement of the decimal point, and making connections to a written method.

139

EUREKA MATH™

5. You can look up information on the world's tallest buildings at
    http://www.infoplease.com/ipa/A0001338.html.

    a. The Aon Centre in Chicago, Illinois, is one of the world's tallest buildings. Built in 1973, it is 1,136 feet
        high and has 80 stories. If each story is of equal height, how tall is each story?

    b. Burj al Arab Hotel, another one of the world's tallest buildings, was finished in 1999. Located in
        Dubai, it is 1,053 feet high with 60 stories. If each floor is the same height, how much taller or shorter
        is each floor than the height of the floors in the Aon Center?

Lesson 26:   Divide decimal dividends by two-digit divisors, estimating quotients,
                  reasoning about the placement of the decimal point, and making
                  connections to a written method.

EUREKA
MATH™

©2015 Great Minds. eureka-math.org
G5-M2-SE-B1-1.3.1-01.2016

Name _____   Date _____

1. Divide. Check your work with multiplication.

a. 5.6 ÷ 16

b. 21 ÷ 14

c. 24 ÷ 48

d. 36 ÷ 24

e. 81 ÷ 54

f. 15.6 ÷ 15

g. 5.4 ÷ 15

h. 16.12 ÷ 52

i. 2.8 ÷ 16

Lesson 27:   Divide decimal dividends by two-digit divisors, estimating quotients, reasoning about the placement of the decimal point, and making connections to a written method.

141

2. 30.48 kg of beef was placed into 24 packages of equal weight. What is the weight of one package of beef?

3. What is the length of a rectangle whose width is 17 inches and whose area is 582.25 in^2?

Lesson 27:   Divide decimal dividends by two-digit divisors, estimating quotients, reasoning about the placement of the decimal point, and making connections to a written method.

EUREKA MATH

4. A soccer coach spent $162 dollars on 24 pairs of socks for his players. How much did five pairs of socks cost?

5. A craft club makes 95 identical paperweights to sell. They collect $230.85 from selling all the paperweights. If the profit the club collects on each paperweight is two times as much as the cost to make each one, what does it cost the club to make each paperweight?

Lesson 27: Divide decimal dividends by two-digit divisors, estimating quotients, reasoning about the placement of the decimal point, and making connections to a written method.

143

Name _____   Date _____

1. Divide. Check your work with multiplication.

    a. $7 \div 28$ b. $51 \div 25$ c. $6.5 \div 13$

    d. $132.16 \div 16$ e. $561.68 \div 28$ f. $604.8 \div 36$

2. In a science class, students water a plant with the same amount of water each day for 28 consecutive days. If the students use a total of 23.8 liters of water over the 28 days, how many liters of water did they use each day? How many milliliters did they use each day?

Lesson 27:   Divide decimal dividends by two-digit divisors, estimating quotients, reasoning about the placement of the decimal point, and making connections to a written method.

EUREKA MATH™

3. A seamstress has a piece of cloth that is 3 yards long. She cuts it into shorter lengths of 16 inches each.
   How many of the shorter pieces can she cut?

4. Jenny filled 12 pitchers with an equal amount of lemonade in each. The total amount of lemonade in the
   12 pitchers was 41.4 liters. How many liters of lemonade would be in 7 pitchers?

Lesson 27:   Divide decimal dividends by two-digit divisors, estimating quotients,   145
              reasoning about the placement of the decimal point, and making
              connections to a written method.

©2015 Great Minds. eureka-math.org
G5-M2-SE-B1-1.3.1-01.2016

This page intentionally left blank

Name _____   Date _____

1. Ava is saving for a new computer that costs $1,218. She has already saved half of the money. Ava earns $14.00 per hour. How many hours must Ava work in order to save the rest of the money?

2. Michael has a collection of 1,404 sports cards. He hopes to sell the collection in packs of 36 cards and make $633.75 when all the packs are sold. If each pack is priced the same, how much should Michael charge per pack?

EUREKA
MATH™

Lesson 28:   Solve division word problems involving multi-digit division with group
             size unknown and the number of groups unknown.

147

©2015 Great Minds. eureka-math.org
G5-M2-SE-B1-1.3.1-01.2016

3. Jim Nasium is building a tree house for his two daughters. He cuts 12 pieces of wood from a board that is 128 inches long. He cuts 5 pieces that measure 15.75 inches each and 7 pieces evenly cut from what is left. Jim calculates that, due to the width of his cutting blade, he will lose a total of 2 inches of wood after making all of the cuts. What is the length of each of the seven pieces?

4. A load of bricks is twice as heavy as a load of sticks. The total weight of 4 loads of bricks and 4 loads of sticks is 771 kilograms. What is the total weight of 1 load of bricks and 3 loads of sticks?

Lesson 28:   Solve division word problems involving multi-digit division with group size unknown and the number of groups unknown.

EUREKA MATH

©2015 Great Minds. eureka-math.org
G5-M2-SE-B1-1.3.1-01.2016

Name _____   Date _____

1. Mr. Rice needs to replace the 166.25 ft of edging on the flower beds in his backyard. The edging is sold in lengths of 19 ft each. How many lengths of edging will Mr. Rice need to purchase?

2. Olivia is making granola bars. She will use 17.9 ounces of pistachios, 12.6 ounces of almonds, 12.5 ounces of walnuts, and 12.5 ounces of cashews. This amount makes 25 bars. How many ounces of nuts are in each granola bar?

Lesson 28:   Solve division word problems involving multi-digit division with group size unknown and the number of groups unknown.

149

©2015 Great Minds. eureka-math.org
G5-M2-SE-B1-1.3.1-01.2016

3. Adam has 16.45 kg of flour, and he uses 6.4 kg to make hot cross buns. The remaining flour is exactly enough to make 15 batches of scones. How much flour, in kg, will be in each batch of scones?

4. There are 90 fifth-grade students going on a field trip. Each student gives the teacher $9.25 to cover admission to the theater and for lunch. Admission for all of the students will cost $315, and each student will get an equal amount to spend on lunch. How much will each fifth grader get to spend on lunch?

Lesson 28:   Solve division word problems involving multi-digit division with group
                                 size unknown and the number of groups unknown.

©2015 Great Minds. eureka-math.org
G5-M2-SE-B1-1.3.1-01.2016

5. Ben is making math manipulatives to sell. He wants to make at least $450. Each manipulative costs $18 to make. He is selling them for $30 each. What is the minimum number he can sell to reach his goal?

Lesson 28: Solve division word problems involving multi-digit division with group size unknown and the number of groups unknown.

151

©2015 Great Minds. eureka-math.org
G5-M2-SE-B1-1.3.1-01.2016

This page intentionally left blank

Name _____   Date _____

Solve.

1. Lamar has 1,354.5 kilograms of potatoes to deliver equally to 18 stores. 12 of the stores are in the Bronx. How many kilograms of potatoes will be delivered to stores in the Bronx?

2. Valerie uses 12 fluid oz of detergent each week for her laundry. If there are 75 fluid oz of detergent in the bottle, in how many weeks will she need to buy a new bottle of detergent? Explain how you know.

Lesson 29:   Solve division word problems involving multi-digit division with group size unknown and the number of groups unknown.   153

©2015 Great Minds. eureka-math.org
G5-M2-SE-B1-1.3.1-01.2016

3. The area of a rectangle is 56.96 m². If the length is 16 m, what is its perimeter?

4. A city block is 3 times as long as it is wide. If the distance around the block is 0.48 kilometers, what is the area of the block in square meters?

Solve division word problems involving multi-digit division with group size unknown and the number of groups unknown.

Name _____   Date _____

Solve.

1. Michelle wants to save $150 for a trip to the Six Flags amusement park. If she saves $12 each week, how many weeks will it take her to save enough money for the trip?

2. Karen works for 85 hours throughout a two-week period. She earns $1,891.25 throughout this period. How much does Karen earn for 8 hours of work?

Lesson 29:   Solve division word problems involving multi-digit division with group   **155**
                    size unknown and the number of groups unknown.

©2015 Great Minds. eureka-math.org
G5-M2-SE-B1-1.3.1-01.2016

3. The area of a rectangle is 256.5 m². If the length is 18 m, what is the perimeter of the rectangle?

4. Tyler baked 702 cookies. He sold them in boxes of 18. After selling all of the boxes of cookies for the same amount each, he earned $136.50. What was the cost of one box of cookies?

Lesson 29:   Solve division word problems involving multi-digit division with group size unknown and the number of groups unknown.

©2015 Great Minds. eureka-math.org
G5-M2-SE-B1-1.3.1-01.2016

5. A park is 4 times as long as it is wide. If the distance around the park is 12.5 kilometers, what is the area of the park?

Lesson 29:   Solve division word problems involving multi-digit division with group   157
              size unknown and the number of groups unknown.

©2015 Great Minds. eureka-math.org
G5-M2-SE-B1-1.3.1-01.2016

This page intentionally left blank

This page intentionally left blank

This page intentionally left blank

This page intentionally left blank

This page intentionally left blank

This page intentionally left blank

This page intentionally left blank